CENSURE AND SANCTIONS

Censure and Sanctions

Andrew von Hirsch

CLARENDON PRESS · OXFORD
1993

Oxford University Press, Walton Street, Oxford OX2 6DP
Oxford New York Toronto
Delhi Bombay Calcutta Madras Karachi
Kuala Lumpur Singapore Hong Kong Tokyo
Nairobi Dar es Salaam Cape Town
Melbourne Auckland Madrid
and associated companies in
Berlin Ibadan

Oxford is a trade mark of Oxford University Press

Published in the United States
by Oxford University Press Inc., New York

British Library Cataloguing in Publication Data
Data available

Library of Congress Cataloging in Publication Data
von Hirsch, Andrew.
Censure and sanctions/Andrew von Hirsch.
p.   cm.
Includes bibliographical references.
1. Sentences (Criminal procedure)—Great Britain.   2. Sentences
(Criminal procedure)   I. Title.
KD8406.V66   1993
345.41'0772—dc20
[344.105772]   93–4428
ISBN 0–19–825768–6

1 3 5 7 9 10 8 6 4 2

Typeset by Cambrian Typesetters
Frimley, Surrey
Printed in Great Britain
on acid-free paper by
Bookcraft Ltd., Midsomer Norton, Bath

For my son
Alexander

General Editor's Introduction

WORKS on theories of punishment are legion. This one breaks new ground by tackling, from a principled point of view, a number of contemporary issues in penal policy. The basis for the arguments is provided by the re-statement of the foundations of desert theory in the second chapter, and subsequent chapters draw upon other penological and philosophical perspectives in grappling with issues such as the overall level of punitiveness, the use of community sanctions, the intrusiveness of penal measures, and so forth. The author also brings to these topics his wide understanding of sentencing issues in Europe and North America, as is particularly evident in Chapter 10. At a time when discussion of the sentencing of offenders so often becomes smothered in political rhetoric, it is all the more important to make advances on the front of principle. I am delighted that Andrew von Hirsch has written a book which does just that.

Andrew Ashworth

Preface

I am somewhat surprised to have written this book. I wrote two works on sentencing theory, one published in 1976 and the other in 1985. That seemed enough, and I was beginning work on other topics.

Questions of punishing convicted offenders—particularly, those regarding the principle of proportionate sanctions—kept pressing themselves on my attention, however. New philosophical writing on censure and its moral functions was appearing, and set me thinking about how censure, punishment, and proportionality are linked. It was becoming ever more apparent that proportionalist sentencing theory needed, but did not have, an explicit conception of how a penalty scale should be anchored. Experimentation with 'intermediate' non-custodial sanctions was also beginning, raising questions about how such sanctions should be allocated. The development of those sanctions also raised disturbing collateral questions—for example, of how degrading penal routines (modern versions of the pillory or the stocks) could be prevented. I wrote articles on some of these issues in various journals.

The enactment of England's Criminal Justice Act of 1991 became the immediate stimulus for writing the book. Here, a large and influential jurisdiction had approved legislation making proportionality the primary criterion for deciding penalties—for determining not only the use of prison terms, but of non-custodial sanctions as well. Rather than requiring judges to apply a table of prescribed sentences, the Act calls upon them to interpret the idea of proportionality. It is too early to see whether the Act succeeds in its aim of promoting proportionate sanctions, but it has surely evoked extensive interest. I began to feel that a book on the still-unresolved issues of proportionality was timely.

I first thought that the volume might consist of the articles I had written, plus some additional chapters. But as I looked these materials over, I found them insufficient. A new book was needed. As I started work, a variety of questions presented themeselves to me that I had not considered before. The result of this rethinking process is the present work.

Punishment is a fascinating but uncomfortable question. Having the State deliberately impose pain on convicted offenders can be no better than a sad necessity. It is not easy to conceive of how to make such a system humane and fair—or even, not too inhumane and unfair. Maintaining a tolerable system of sanctions in practice is harder still, as politics (the topic at the end of this book) inevitably intrudes. No one should delude themselves that there are neat solutions.

Acknowledgements

A number of people helped me conceptualize this book. Andrew Ashworth, the general editor of the series in which this volume appears, helped me to define the project, co-authored one of the chapters, and provided me with detailed and helpful commentary. Uma Narayan co-authored another chapter, and helped me think through a number of philosophical issues. John Kleinig gave me valuable critical advice on how the two elements which I think justify punishment's existence, censure and prevention, might be linked. Don Scheid asked some keen sceptical questions about censure's justifying role—ones which I hope, but am far from certain, I have answered. Tatjana Hoernle pressed me on questions of anchoring the penalty scale.

A number of other colleagues and friends reviewed the first full draft of the book, or particular chapters of it, and provided valuable comments: Douglas Husak, Nils Jareborg, Julian Roberts, Paul Robinson, and Michael Tonry.

I am grateful to my reseach assistant, Bruce Taylor, for his painstaking work in reviewing several earlier drafts of the book, and providing substantive as well as technical suggestions.

Once again, I thank Joan Schroeder for typing portions of the manuscript, and helping me with some of the mysteries of wordprocessing. Sandra Wright also assisted in the typing.

In two chapters, I have used (with substantial alterations and additions) portions of articles of mine that have been published elsewhere. Their publishers have kindly given their permission, as follows:

For Chapter 3: 'Not Not Just Deserts: A Response to Braithwaite and Pettit', *Oxford Journal of Legal Studies* 12 (1992), 83–98 (with Andrew Ashworth).

For Chapter 9: 'The Politics of "Just Deserts"', *Canadian Journal of Criminology* 32 (1990), 397–413.

Contents

1

Introduction

IN a field as fashion-ridden as criminology has been, it is surprising that a single idea can preserve its influence for several decades. Yet desert theory has done so: its notion of proportionality of sentence—which came to prominence in the mid-1970s[1]—is still with us today. Testimony to the theory's continuing influence is England's Criminal Justice Act of 1991.[2] The statute, for the first time in that country, sets forth general principles for choosing sentence. The Act's statutory principles rely substantially on the idea of proportionality: that the sentence should be proportionate in its severity to the gravity of the offence.

Why the potency of this idea of the proportionate sentence? First, it is ethically plausible. Most of us, as part of our everyday notions of justice, think that penalties should fairly reflect the degree of blameworthiness of the conduct involved. Even children object when they notice disparities in the punishments they receive for similar acts of misbehaviour. Preventive penal strategies, by contrast, seem ethically troublesome. For example, giving someone extra punishment on the basis of a prediction that he will reoffend looks like penalizing him for an offence he is yet to commit.

Second, the idea of proportionality provides sentencers with a degree of guidance, in a way that competing theories seldom do. What policymakers and judges chiefly need from a sanctioning theory is help in scaling punishments—in deciding whether this type of convicted offender should be penalized more or less than that type, and how much more or how much less. Proportionality provides at least a partial answer: penalties should be graded in severity to reflect the gravity of the offences involved.

Competing utilitarian theories seldom appear capable of providing much guidance at all. The difficulty has been evident from their first formulation. Jeremy Bentham, writing two centuries ago, proposed a calculus according to which penalties would be determined by weighing their deterrent yields against the suffering caused by their imposition.[3] Such a scheme is capable of producing concrete answers only if those costs and deterrent yields are ascertainable, but they almost never are. Little is known, for example, about the marginal deterrent effects of varying the severity of penalties.[4] Replacing deterrence with rehabilitation scarcely improves matters. While debate continues on the extent to which penal treatments can be made to work, no one has been able to show how sentences can *routinely* be decided on the basis of treatment.[5]*

That proportionality has these attractions does not suffice, however, to establish its merits. Even if sufficient knowledge does not now exist to array penalties on preventive grounds, such knowledge might eventually be obtainable. Even if proportionality comports with our intuitive sense of fairness, we still need to inquire *why* it is a requirement of justice. Desert theory also faces challenges—for example, from theorists who urge that proportionality-constraints should be relaxed for the sake of various other objectives.[6] Such objections need to be addressed.

1. THE EXPERIENCE OF SENTENCING REFORM

What has been the experience with sentencing reform in English-speaking countries? The United States of America's active reform era was in the late 1970s and early 1980s.[7] Perhaps the most significant development of that period was the writing of Minnesota's sentencing guidelines. Designed to reflect a modified version of a desert rationale, these standards did achieve a modicum of success in producing more consistent sentence outcomes, and in restraining the growth of prison populations.[8]

In the last half of the 1980s, however, the pace of reform has slowed, and there has developed a sharp politicization of punishment policy. While crime rates have not risen during this period, public impatience about crime seems to have done so—as has surely the willingness of politicians to support draconian sanctions as proof of their determination to combat crime. One symptom of this new mood has been the 'war against drugs'— with its campaign for escalated penalties for drugs related offences at both the state and federal level.[9] Another symptom has been the demise of federal sentencing reform. Early in the decade, a US Sentencing Commission was established, with much fanfare and considerable resources, to deal with cases involving infringements of federal criminal statutes. The Commission's guidelines, issued in 1987, sought to accommodate demands for toughness from the Administration and Congress, by producing a harsh set of standards—designed sharply to increase the use of imprisonment. Those standards also abjure reliance on any particular rationale, and are cumbersome to apply.[10]

There has, however, been one encouraging recent development in the USA: experimentation with non-custodial sanctions of intermediate severity.[11] Some states and localities are beginning to consider guidelines

---

* For a discussion of the empirical and ethical issues associated with reviving the rehabilitative ethic in sentencing, see A. von Hirsch and L. Maher, 'Should Penal Rehabilitationism Be Revived?' (1992).

for these sanctions' use.[12] This raises the question, explored here, of how such penalties may fairly be scaled.

In England, until the early 1980s, there was relatively little interest in sentencing issues. Since then, however, an extensive literature has developed on proportionality of sentence,[13] and on techniques for guiding sentencing discretion.[14] The big surprise came in 1989, when the Government produced a White Paper[15] advocating systematic sentencing reform. The Paper proposed that proportionality should be the primary criterion for deciding both the use of imprisonment and the use of non–custodial penalties. Of particular interest was the fact that the White Paper proposed using such a desert-oriented scheme to *reduce* reliance on imprisonment. A rationale emphasizing the character of the current offence, the report suggested, would make it easier to avoid imprisoning recidivist property offenders.[16]★ The White Paper evolved, with various changes and omissions, into the Criminal Justice Act of 1991.[17]

Whether England's 1991 Act succeeds in its aims depends largely on how the judiciary is prepared to apply it, and the possibility remains of nullification through disregard of the Act's provisions. Nevertheless, the new changes have produced an unprecedented degree of interest in sentencing reform and sanctioning theory.

2. ISSUES ADDRESSED IN THIS BOOK

The developments of the last decade have prompted a new set of questions on the rationale and criteria for proportionate sentencing. Those questions, with which this book deals, include:

Censure and the rationale for proportionality. In earlier works,[18] I relied upon the idea of penal censure to explain why sentences should be proportionate to the gravity of crimes. It is because punishment expresses blame, I argued, that sanctions should comport with the blameworthiness (i.e. seriousness) of the criminal conduct. However, this idea of censure requires more scrutiny. Why should punishment express blame? Has punishment only a reprobative function, or should it have a preventive one as well? If so, how are the two functions related? Just how does penal censure undergird proportionality? There has been some interesting recent philosophical discussion of blaming and its moral functions,[19] to which my analysis will refer.

Anchoring the penalty scale. An oft-heard criticism of desert theory has been that (while it may help to order penalties relative to one another) it says

---

★ In the American political environment, any comparable official suggestion would have been unlikely. Reforms could aim, at most, at restraining further increases, as Minnesota's guidelines were designed to do.

little about how severe or lenient the penalty scale as a whole should be.[20] It is high time that this critical question is addressed. I shall sketch a conception of how the penalty scale should be anchored—one that would call for a substantial reduction in penalty levels.

Hybrid models. Many reformers are attracted by the idea of proportionate punishments, but wish to include other aims as well. In a 1987 essay,[21] Paul Robinson began to explore the idea of 'hybrid' schemes—particularly, those that rely on desert primarily, but with limited deviations permitted to achieve other objectives. The construction and rationale of such hybrids is worth examining further.

Intermediate punishments. A decade ago, imprisonment and probation were seen as the primary penalties. Sentencing guidelines were designed to guide and limit the use of imprisonment. Recent years have been witnessing, as I just noted, a new interest in 'intermediate' sanctions—in measures such as unit or 'day' fines measured by income, community service, home detention, and the like.[22] It is noteworthy that England's 1991 Criminal Justice Act calls for expanded reliance on such intermediate penalties, and applies the proportionality principle to their use.[23] However, there has been relatively little theoretical treatment of the scaling of non-custodial sanctions.[24] The new interest in non-incarcerative penalties raises subsidiary questions also. One concern is to what extent such penalties should be 'individualized', that is, addressed to the situation of the particular actor. The other issue is how to rule out modern versions of the stocks: that is, degrading or intrusive penal routines.

The politics of 'just deserts'. Some critics[25] argue that trying to punish offenders as they deserve leads to harsher punishments. The links between penal purpose and criminal-justice politics are, in my opinion, much more complex than this simple thesis assumes. I shall try to give some closer scrutiny to the politics of proportionality.

3. ASSUMPTIONS OF THIS BOOK

Before proceeding with these issues, let me mention three assumptions on which this work is based.[26] One is that justice matters, and indeed, should have primacy in deciding the allocation of criminal sanctions. Much of this book concerns what is a fair way of allocating penalties.

My second assumption is that parsimony of punishment counts. Punishments hurt those who must undergo them, and a decent society should seek to keep the purposeful infliction of hurt to a minimum. Some writers have tried to define parsimony in utilitarian terms: lesser punishments are to be preferred except where greater ones are warranted for crime prevention. This, obviously, would bias the discussion in favour of utilitarian schemes.[27] Parsimony, correctly understood, should not presuppose a particular set of penal aims. Desert can be applied parsimoniously, provided its criteria are applied so as to scale penalties down. It is this approach that will be defended in the present volume.

My third assumption concerns the nature of offending. Penologists are drawn too easily to viewing offenders and potential offenders as a species apart from law-abiding citizens, as persons who largely are impervious to moral appeals, and who need to be intimidated or restrained into compliance with the law. I do not accept this view. A sanctioning system should not be seen as one which 'we' devise to prevent 'them' from offending. Rather, it should be one which free citizens could devise to regulate their *own* conduct. Persons should be assumed to be both susceptible (at least under some circumstances) to the temptation to offend, and capable of understanding the moral judgements which the criminal sanction conveys. A sanctioning system, in a democratic society, should be of the kind which such persons could accept, as a way of assisting them to resist their own temptations, in a manner that respects their capacity for choice.

2

Censure and Proportionality

THE principle of proportionality—that sanctions be proportionate in their severity to the gravity of offences—appears to be a requirement of justice. People have a sense that punishments which comport with the gravity of offences are more equitable than punishments that do not. However, appeals to intuition are not enough: the principle needs to be supported by explicit reasons. What are those reasons?

Although 'Why Punish Proportionately?' is ultimately an ethical question, it has not been explored much by philosophers. Philosophical writing has chiefly confined itself to the general justification for punishment, why the criminal sanction should exist at all. Seldom addressed, however, has been what bearing the justification for punishment's existence has on the question of how much offenders should be penalized.

It is the last question which will be examined in this chapter. Does sentencing theory change, depending on the general justification for punishment? Does one or another such justification provide support for the principle of proportionality, and why is this so? I shall examine two kinds of desert-based general justifications that have attracted recent philosophical attention. One theory sees the institution of punishment as rectifying the 'unfair advantage' which lawbreakers obtain by offending. The other focuses on punishment's role as expressing censure or reprobation. With each theory, I shall ask whether it supports proportionate sanctions, and how convincingly it does so.

I begin with the 'unfair advantage' theory. A brief analysis will suggest that the theory, apart from its intrinsic perplexities, provides poor support for the principle of proportionality. I turn, then, to censure. I shall defend an account of the criminal sanction that emphasizes its reprobative features, and then explain why this account supports a requirement of proportionate sanctions. In my account, the institution of punishment has preventive as well as reprobative features. This also will require me to explore the relationship between the criminal sanction's censuring and preventive aspects.

Any desert-based theory of legal punishment assumes that criminal conduct is, in some sense, reprehensible. Censure-based theories clearly have this presupposition, for the conduct is treated as warranting blame. Criminal prohibitions of today have wide scope, however, and include conduct that seems in no plausible way blameworthy. A desert-based theory of punishment, however, need not defend all such prohibitions. It suffices if the core conduct with which the criminal law deals—acts of violence or fraud, for example—can reasonably be described as being reprehensible. At issue here is whether *any* conduct should be legally punishable, and if so, how much punishment that conduct should receive. It is not necessary to defend the criminal law in its full present scope.

Is it so clear, however, that even the core conduct addressed by the criminal law is blameworthy? Conceivably, the State is ill-situated to make authoritative moral judgements at all. Perhaps, the limited life-options of many criminals puts their culpability in doubt. The present chapter, however, is complex enough without introducing these issues, and so I will postpone discussing them until the end of the book.

1. THE 'UNFAIR-ADVANTAGE' THEORY

The unfair-advantage (or 'benefits-and-burdens') theory has been attributed to Kant, but whether Kant actually subscribed to it is debatable.[1] The first unequivocal statement of this position appeared two decades ago in the writings of Herbert Morris and Jeffrie Murphy, although both authors have distanced themselves from it recently.[2] A number of other contemporary philosophers, however, continue to defend the theory.[3]*

The unfair-advantage view offers a retributive, retrospectively-oriented account of why offenders should be made to suffer. The account focuses on the criminal law as a jointly beneficial enterprise. The law requires each person to desist from certain kinds of predatory conduct. By so desisting, the person benefits others; but he also benefits from their reciprocal self-restraint. The person who victimizes others while benefiting from their self-restraint thus obtains an unjust advantage. Punishment's function is to impose an offsetting disadvantage.

This theory has various perplexities.[4]** It is arguable (although still debatable***) that the offender, by benefiting from others' self-restraint, has a reciprocal obligation to restrain himself. It is much more obscure, however, to assert that—if he disregards that obligation and does offend— the unfair advantage he supposedly thereby gains can somehow (in other

---

* These include John Finnis, Alan Gewirth, George Sher, and Wojciech Sadurski.

** I did endorse the theory in my 1976 book, *Doing Justice*, as a partial justification for the existence of punishment. But I since have been convinced of its deficiencies, and argued against it already in a 1985 volume.

*** A person's receiving benefits from others does not necessarily put him under a duty to reciprocate, unless he accedes to or accepts those benefits. See, A. J. Simmons, *Moral Principles and Political Obligation* (1979). Appealing to the benefits received from others' self restraint would also be a rather roundabout way of grounding the duty not to offend. It would seem simpler and more plausible to speak of a direct duty not to infringe the rights of others.

than a purely metaphorical sense) be eliminated or cancelled by punishing him. In what sense does his being deprived of rights *now* offset the extra freedom he has arrogated to himself *then* by offending? And why is preserving the balance of supposed advantages a reason for invoking the coercive powers of the State?

Even if such queries could be answered, the benefits-and-burdens theory has another difficulty: it provides little or no assistance for determining the quantum of punishment. One problem is that the theory would distort the way that the gravity of crimes is assessed. R. A. Duff has pointed out the artificiality of describing typical victimizing crimes, such as armed robbery, in terms of the advantage the robber gains over uninvolved third parties, rather than in terms of the conduct's intrusion into the rights of victims.[5] Certain types of offence, it is true, might plausibly be explained in terms of unjustified advantage. Tax evasion is an example: it seems to involve taking more than one's share. Although the tax evader refuses to pay his own tax, he gets the benefit of others' payments through the services he receives. Tax evasion, however, is scarcely the paradigm criminal offence, and it is straining to try to explain the heinousness of common crimes such as burglary and robbery in similar fashion.

The theory also provides little or no intelligible guidance on how much punishment an offence of any given degree of seriousness should receive. It is not concerned with literal advantage or disadvantage: what matters, instead, is the additional freedom of action that the offender has unfairly appropriated. But the notion of degrees of freedom is not helpful in making comparisons among crimes. It is one thing to say that the armed robber or the burglar permits himself actions that others refrain from taking, and thereby unfairly gains a liberty that others have relinquished in their (and his) mutual interest. It is different, and much more opaque, to say the robber deserves more punishment than the burglar because he somehow has arrogated to himself a greater degree of unwarranted freedom *vis-à-vis* others.[6]*

---

* Sadurski has asserted that the extent of the offender's 'benefit' from not having exercised self-restraint varies with the importance of the rights infringed. However, he does not offer a convincing account of why violating a more 'important' prohibition involves taking a greater degree of unwarranted freedom. Why is one more 'free' if one takes another person's life than his property?

Michael Davis claims to have devised a proxy for the offender's 'advantage': the price at which a licence for the behaviour might be had. He asks us to imagine a society in which the government auctioned licences to commit limited numbers of offences of various sorts. Potential offenders would bid, and the size of the bids would reflect the bidder's estimate of the value to himself of having the freedom to engage in the conduct. The rankings of bids resulting would be used to decide the comparative severities of sanctions.

I have criticized Davis's model at some length elsewhere. Suffice it to say here that the licence analogy is misplaced. A licence is permissive: one may legitimately engage in the conduct if one pays the fee. The criminal law, however, is prohibitive and condemnatory: one ought not to engage in the conduct even if one is willing to 'pay the price' by suffering the sanction. It would be surprising if licences illuminated the issue of allocating punishments.

2. CENSURE-BASED JUSTIFICATIONS FOR PUNISHMENT

Reprobative accounts of the institution of the criminal sanction are those that focus on that institution's condemnatory features, that is, its role as conveying censure or blame. The penal sanction clearly does convey blame. Punishing someone consists of visiting a deprivation (hard treatment) on him, because he supposedly has committed a wrong, in a manner that expresses disapprobation of the person for his conduct. Treating the offender as a wrongdoer, Richard Wasserstrom has pointed out,[7] is central to the idea of punishment. The difference between a tax and a fine does not rest in the kind of material deprivation (money in both cases). It consists, rather, in the fact that the fine conveys disapproval or censure, whereas the tax does not.[8]

An account of the criminal sanction which emphasizes its reprobative function has the attraction of being more comprehensible, for blaming is something we do in everyday moral judgements. A censure-based account is also easier to link to proportionality: if punishment conveys blame, it would seem logical that the quantum of punishment should bear a reasonable relation to the degree of blameworthiness of the criminal conduct.

Why the Censure?

That punishment conveys blame or reprobation is, as just mentioned, evident enough. But why *should* there be a reprobative response to the core conduct with which the criminal law deals? Without an answer to that question, legal punishment might arguably be replaced by some other institution that has no blaming implications—a response akin to a tax meant to discourage certain behaviour.

P. F. Strawson provides the most straightforward account.[9] The capacity to respond to wrongdoing by reprobation or censure, he says, is simply part of a morality that holds people accountable for their conduct. When a person commits a misdeed, others judge him adversely, because his conduct was reprehensible. Censure consists of the expression of that judgement, plus its accompanying sentiment of disapproval. It is addressed to the actor because he or she is the person responsible. One would withhold the expression of blame only if there were special reasons for not confronting the actor: for example, doubts about one's standing to challenge him.

Davis's model, moreover, fails to reflect the logic of the unfair-advantage theory. Whereas the theory addresses the additional freedom of action the offender arrogates to himself by offending, the bids at the hypothetical auction would depend on *literal* advantage: if a large theft is more profitable than a killing, it will attract the higher bid.

While Strawson's account seems correct as far as it goes—blaming *does* seem part of holding people accountable for their actions—it may be possible to go a bit further and specify some of the positive moral functions of blaming.

Censure addresses the victim. He or she has not only been injured, but *wronged* through someone's culpable act. It thus would not suffice just to acknowledge that the injury has occurred or convey sympathy (as would be appropriate when somone has been hurt by a natural catastrophe). Censure, by directing disapprobation at the person responsible, acknowledges that the victim's hurt occurred through another's fault.[10]

Censure also addresses the act's perpetrator. He is conveyed a certain message concerning his wrongful conduct, namely that he culpably has injured someone, and is disapproved of for having done so. Some kind of moral response is expected on his part—an expression of concern, an acknowledgement of wrongdoing, or an effort at better self-restraint. A reaction of indifference would, if the censure is justified, itself be grounds for criticizing him.*

Censure gives the actor the opportunity for so responding, but it is not a technique for evoking specified sentiments. Were inducing penitent reflection the chief aim, as R.A. Duff has claimed,[11] there would be no point in censuring actors who are either repentant or defiant. The repentant actor understands and regrets his wrongdoing already; the defiant actor will not accept the judgement of disapproval which the censure expresses.[12] Yet we would not wish to exempt from blame either the repentant or the seemingly incorrigible actor. Both remain moral agents, capable of understanding others' assessment of their conduct—and censure conveys that assessment. The repentant actor finds his self-evaluation confirmed through the disapproval of others; the defiant actor is made to feel and understand the disapproval of others, whatever he himself may think of his conduct. Such communication of judgement and feeling is the essence of moral discourse among rational agents.

Were the primary aim that of producing actual changes in the actor's moral attitudes, moreover, the condemnor would ordinarily seek information about his personality and outlook, so as better to foster the requisite attitudinal changes. But blaming, in ordinary life as well as in more formal contexts, does not involve such enquiries. One ascribes wrongdoing to the actor and conveys the disapprobation—limiting enquiry about the actor to questions of his capacity for choice. The condemnor's role is not that of the mentor or priest.**

The criminal law gives the censure it expresses yet another role: that of addressing third parties, and providing them with reason for desistence.

---

* For a useful further analysis of these functions, see Uma Narayan, 'Adequate Responses and Preventive Benefits' (1993).
** For fuller development of this point, see Ch. 8.

Unlike blame in everyday contexts, the criminal sanction announces in advance that specified categories of conduct are punishable. Because the prescribed sanction is one which expresses blame, this conveys the message that the conduct is reprehensible, and should be eschewed. It is not necessarily a matter of inculcating that the conduct is wrong, for those addressed (or many of them) may well understand that already. Rather, the censure embodied in the prescribed sanction serves to *appeal* to people's sense of the conduct's wrongfulness, as a reason for desistence.[13]★

This normative message expressed in penal statutes is not reducible, as penal utilitarians might suppose, to a mere inducement to compliance—one utilized because the citizenry could be more responsive to moral appeals than bare threats. If persons are called upon to desist because the conduct is wrong, there ought to be good reasons for supposing that it *is* wrong; and the message expressed through the penalty about its degree of wrongfulness ought to reflect how reprehensible the conduct indeed is. This point will be elaborated upon in the next chapter, where penal censure is contrasted with instrumentalist strategies of 'shaming'.

The foregoing account explains why predatory conduct should not be dealt with through neutral sanctions that convey no disapproval. Such sanctions—even if they were no less effective in discouraging the behaviour—deny the status of the person as an agent capable of moral understanding. A neutral sanction would treat offenders or potential offenders much as tigers might be treated in a circus, as beings that have to be restrained, intimidated, or conditioned into compliance because they are incapable of understanding why biting people (or other tigers) is wrong. A condemnatory sanction treats the actor as a *person* who is capable of such understanding.

A committed utilitarian might insist that treating the actor as a person in this fashion can only be warranted on instrumental grounds. Nigel Walker takes this view: if '. . . the message which expresses blame need have no utility,' he asserts, 'where lies the moral necessity?'[14] This, however, is reductionist. Treating the actor as someone capable of choice, rather than as a tiger, is a matter of acknowledging his dignity as a human being. Is this acknowledgement warranted only if it leads to beneficial social consequences? Those consequences would not necessarily be those of crime prevention—for, as just noted, it might be possible to devise a 'neutral' sanction (one designed to visit material deprivation but convey no blame) that prevents crime at least as well. Might society somehow have better cohesion if actors are treated as being responsible (and hence subject to

---

★ I thus would not subscribe to views that treat penal censure as primarily a matter of moral education, that is, of inculcating standards. For a fuller critique of such views (including those of Herbert Morris and Jean Hampton), see Narayan, 'Moral Education and Criminal Punishment' (1993).

censure) for their actions? Making such a claim would involve trying to reduce ethical judgments to difficult-to-confirm predictions about social structure. While one can have some confidence in the moral judgement that offenders should be treated as agents capable of choice, it will be difficult to verify that so treating them will lead to a more smoothly-running society.

Why the Hard Treatment?

It is still necessary to address punishment's other constitutive element: deprivation or hard treatment. Some desert theorists (John Kleinig and Igor Primoratz,[15] for example) assert that notions of censure can account also for the hard treatment. They argue that censure (at least in certain social contexts) cannot be expressed adequately in purely verbal or symbolic terms; that hard treatment is needed to show that the disapprobation is meant seriously. For example, an academic department does not show disapproval of a serious lapse by a colleague merely through a verbal admonition; to convey the requisite disapproval, some curtailment of privileges is called for. This justification has plausibility outside legal contexts, where the deprivations involved are modest enough to serve chiefly to underline the intended disapproval. However, I doubt that the argument sustains the criminal sanction.

The criminal law seems to have preventive features in its very design. When the State criminalizes conduct, it issues a legal threat: such conduct is proscribed, and violation will result in the imposition of specified sanctions. The threat appears to be explicitly aimed at discouraging the proscribed conduct.[16] Criminal sanctions also seem too onerous to serve just to give credibility to the censure. Even were penalties substantially scaled down from what they are today, some of them still could involve significant deprivations of liberty or property. In the absence of a preventive purpose, it is hard to conceive of such intrusions as having the sole function of showing that the State's disapproval is seriously intended.

This reasoning led me to suggest, in a 1985 volume,[17] a bifurcated account of punishment. The penal law, I said, performs two interlocking functions. By threatening unpleasant consequences, it seeks to discourage criminal behaviour. Through the censure expressed by such sanctions, the law registers disapprobation of the behaviour. Citizens are thus provided with moral and not just prudential reasons for desistence.

However, the two elements in my account, reprobation and prevention, remained uneasily matched. Whereas the censuring element appeals to the person's moral agency, does not the preventive element play merely on his fear of unpleasant consequences? If the person is capable of being moved by moral appeal, why the threat? If not capable and thus in need of the threat, it appears that he is being treated like a tiger. A clarification of the preventive function—and its relation to the censuring function—is needed.

The preventive function of the sanction should be seen, I think, as supplying a prudential reason that is tied to, and supplements, the normative reason conveyed by penal censure. The criminal law, through the censure embodied in its prescribed sanctions, conveys that the conduct is wrong, and a moral agent thus is given grounds for desistence. He may (given human fallibility) be tempted nevertheless. What the prudential disincentive can do is to provide him a further reason—a prudential one—for resisting the temptation.* Indeed, an agent who has accepted the sanction's message that he ought not offend, and who recognizes his susceptibility to temptation, could favour the existence of such a prudential disincentive, as an aid to carrying out what he himself recognizes as the proper course of conduct.

A certain conception of human nature, of which I spoke in the previous chapter, underlies this idea of the preventive function as a supplementary prudential disincentive. Persons are assumed to be moral agents, capable of taking seriously the message conveyed through the sanction, that the conduct is reprehensible. They are fallible, nevertheless, and thus face temptation. The function of the disincentive is to provide a prudential reason for resisting the temptation. The account would make no sense were human beings much better or worse: an angel would require no appeals to prudence, and a brute could not be appealed to through censure.

Notice that I am speaking here of why prevention might in principle be a legitimate supporting reason for punishment's existence. I have not been speaking of sanctions' severity, where coerciveness could still be a problem. If penalty levels rise too high, the normative reasons for desistence supplied by penal censure could become largely immaterial; and the disincentive become much more than supplementary to the censuring message. If minimal prudence virtually would demand compliance, what difference could the sanction's normative message make? I shall return to this issue of sanction levels, when I discuss a penalty scale's anchoring points in Chapter 5.

The Relation Between the Two Elements

What is the relation between the two elements in punishment, the reprobative and preventive? We need to be careful that the latter does not operate independently, or else we may undermine the proportionality requirement.

Prevention, on the account of it that I have just given, cannot stand alone. If the sanction conveys blame, it may also supply the prudential disincentive

---

* For a comparable analysis of the relation between censure and prevention, and of prevention as a 'supplementary prudential disincentive', see Narayan, 'Adequate Responses to Preventive Benefits'.

I have described—the means for overcoming the temptation. But if it *merely* imposes hard treatment, it remains tiger-control. Granted, a morally committed person might find that even a neutral, non-condemnatory sanction makes it easier to resist temptation and thus comply with a moral obligation he himself recognizes. The sanction itself would not be respectful of his agency, however, if it was couched as a naked demand. Whatever the actor's reasons for compliance, the sanctioner would be treating him as a creature to be controlled, not as someone whose normative reasons for acting matter.

The structure of my proposed justification for punishing is also one in which the blaming function has primacy. A condemnatory response to injurious conduct, I have been arguing, can be expressed either in a purely (or primarily) symbolic mode; or else, in one in which the reprobation is expressed through the visitation of hard treatment. The criminal sanction is a response of the latter kind. It is preferred to the purely symbolic response because of its supplementary role as a disincentive. The preventive function thus operates only *within* a censuring framework.

The censure and the hard treatment are intertwined in the way punishment is structured. A penal measure provides that a specified type of conduct is punishable by certain onerous consequences. Those consequences both constitute the hard treatment and express the reprobation. Altering those consequences—by raising or lowering the penalty on the scale—will alter the degree of censure conveyed. This intertwining of punishment's blaming and hard-treatment features is important for the rationale for proportionality, as we will see.

My two-pronged justification would permit the abolition of the institution of punishment were it not needed for preventive purposes. Imagine a jurisdiction in which social and economic conditions improved so much that predatory conduct became quite rare. The criminal sanction—with its armamentarium of courts, correctional agencies, and sanctions—would cease to be required in order to keep such conduct at tolerable levels. Would such a society still have to preserve this institution to deal with the occasional predatory act? I would think not. The society might wish to maintain some form of official censure to convey the requisite disapproval of such acts, but with the need for prevention eliminated, there would no longer be need for so ambitious, intrusive, and burdensome an institution as the criminal sanction.★

★ The argument for abolition is not quite so simple as the text suggests. Suppose the crime rate has fallen drastically, so that crimes (or at least, those of a substantial nature) become rather rare. Imagine a discussion among a group of citizens about whether punishment should be abolished. Some might argue that such conduct, even if rare, is reprehensible and injurious when it occurs. Even if only a few might be tempted, why not have a sanction that includes a prudential disincentive—to help offset the temptation?

The abolitionists' response would have to be that there are strong countervailing reasons for abolition. The criminal sanction, they might argue, has serious collateral drawbacks: it is prone

3. THE RATIONALE FOR PROPORTIONALITY

So much, then, for the general justification for punishment. It is time to move from 'why punish?' to 'how much?' Assuming a reprobative account of punishment's existence, how can the principle of proportionality be accounted for? The argument will reflect the idea that, if censure conveys blame, its amount should reflect the blameworthiness of the conduct; but it needs to be unpacked more carefully.

Stated schematically, the argument for proportionality involves the following three steps:

1. The State's sanctions against proscribed conduct should take a punitive form; that is, visit deprivations in a manner that expresses censure or blame.
2. The severity of a sanction expresses the stringency of the blame.
3. Hence, punitive sanctions should be arrayed according to the degree of blameworthiness (i.e. seriousness) of the conduct.

Let us examine each of these steps. Step (1) reflects the claim made in the preceding pages: the response to harmful conduct with which the criminal law centrally deals should convey censure. A morally-neutral sanction would not merely be a (possibly) less efficient preventive device; it would be objectionable on the ethical ground that it does not recognize the wrongfulness of the conduct, and does not treat the actor as a moral agent answerable for his or her behaviour.

Step (2) has also been touched upon: in punishment, deprivation or hard treatment is the vehicle for expressing condemnation. When a given type of conduct is visited with comparatively more hard treatment, that signifies a greater degree of disapprobation.[18]*

Step (3)—the conclusion—embodies the claim of fairness. When persons are (and should be) dealt with in a manner ascribing demerit, their treatment should reflect how unmeritorious their conduct can reasonably be said to be. By punishing one kind of conduct more severely than another, the punisher conveys the message that it is worse—which is appropriate only if

---

to errors and abuse and, once established, can lead to unwarranted increases in penalty levels. These drawbacks may have to be tolerated when the criminal sanction assists a significant number of offenders to overcome the temptation to offend. But if such temptation is rare, better let the offender rely on his conscience and whatever form of official censure exists, than create so troublesome an institution as the criminal sanction. (For an analogous argument, see D. Husak, 'Why Punish the Deserving?' (1992).)

* A recent work, by John Braithwaite and Phillip Pettit, attempts to question this link between the quantum of the sanction and the degree of blame. A reply to their arguments is set forth in the next chapter.

the conduct is indeed worse (i.e. more serious). Were penalties ordered in severity inconsistently with the comparative seriousness of crime, the less reprehensible conduct would, undeservedly, receive the greater reprobation.

The foregoing case for proportionality holds if my bifurcated justification for the criminal sanction's existence is adopted. It is not necessary to assert that punishment serves *solely* to express reprobation. In order for my three-step argument to work, it is necessary merely for its premise (Step (1)) to obtain: that the sanction should express reprobation. On my bifurcated view, it should, for I have been arguing why censuring is an essential (albeit not the exclusive) function of the institution of punishment.[19]*

Does my bifurcated account of punishment, however, create a Trojan Horse? If punishment's existence is justified even in part on preventive grounds, might prevention be invoked in deciding comparative severities of punishment? Were that permissible, proportionality would be undermind.[20]**

---

* Even were the institution of punishment justified wholly on preventive grounds, it still expresses blame—and hence (for the reasons explained in Steps (2) and 3)) its allocation criteria should, in fairness, reflect the comparative blameworthiness of offences.

Why, then, not adopt a straightforwardly consequentialist account of a blaming sanction's existence? Such an explanation is readily conceived—indeed, some European theorists have suggested it. In their view, general prevention operates chiefly through the criminal sanction's 'moral-educational' effects, stigmatizing predatory conduct and thereby strengthening citizens' moral inhibitions and making them more reluctant to offend. This stigmatizing effect is achieved through the censure which punishment conveys. Why not be satisfied with such an explanation, if it suffices to support my three-step argument for proportionality? Why bother with all the theorizing of the earlier pages of this chapter concerning the moral basis for penal censure?

The trouble with such an account is that it leaves open an escape hatch. Perhaps these European theorists are right that the criminal sanction, as presently constituted, achieves part of its preventive impact through its 'moral-educational' message. But why not replace that sanction with a 'neutral' one that visits hard treatment but no censure? Such a response might also achieve some prevention, through its purely deterrent or incapacitative effects. Since such a sanction would not express blame, our censure-based argument for proportionality would no longer hold, and the sanction could be distributed without regard to the blameworthiness of the conduct.

My two-pronged general justification—which treats reprobation as a necessary but not sufficient reason for punishment's existence—closes off this escape route; and does so as surely as would wholly reprobative accounts such as Kleinig's or Primoratz's. This is so because my theory would not permit the creation of a neutral scheme of sanctions that purports to visit hard treatment only, for such an institution would fail to recognize the reprehensibleness of the proscribed behaviour.

** David Dolinko makes this objection in a recent article: if punishment has the twin objectives of censure and prevention, why not distribute punishments according to the latter aim? He then notes my reply: that reprobation is not merely an aim but an essential characteristic of punishment—so the comparative severity of the penalty will convey the degree of reprobation. His response is a strange one: tort liability, he says, in some sense conveys censure, since the prerequisite of liability is fault on the part of the actor. Yet the amount of civil recovery depends not on fault but on what is required to make the plaintiff whole. Why, then, need fault be the measure for the quanta of punishments? The answer to Dolinko's contention should be obvious: civil remedies are designed to compensate, and do not have blaming as a central, defining feature. It happens to be that, under existing tort law, fault

Relying on prevention to decide comparative severities is ruled out by something we have mentioned already: the intertwining of punishment's reprobative and hard-treatment features. It is the threatened penal deprivation that expresses the censure as well as serving as the prudential disincentive. Varying the relative amount of the deprivation thus will vary the degree of censure conveyed. Consider a proposal to increase sanctions for a specified type of conduct (beyond the quantum that would be proportionate) in order to create a stronger inducement not to offend. Could such a step be justified under my theory of punishment—on grounds that prevention is said to be part of the general aim of punishing and that this measure achieves prevention more efficiently? No, it could not; thus:

1. Suppose the increase were accomplished simply by raising the prescribed penalty for this type of crime. That increase in punishment would express increased disapprobation for conduct that, *ex hypothesi*, has become no more reprehensible. The increase thus would be objectionable because it treats the offender as more to blame than his conduct warrants.

2. Alternatively, the increase might be accomplished by visiting the proportionate punishment, and then imposing a separate *non-condemnatory* sanction.[21] Since the additional imposition would not be reprobative in character, it would involve no unjustifiable increase in blame. There is, however, another objection: this separate non-condemnatory sanction clearly falls outside my proposed justification for the hard treatment. We are no longer speaking of a censure-expressing response that, for preventive reasons, also involves material deprivation. Instead, the additional sanction is purely preventive and not reprobative at all. It is of the 'tiger-controlling' kind that does not address the actor as a moral agent.

These scenarios confirm what should be apparent, anyway: that making prevention part of the justification for punishment's existence, in the manner that I have, does not permit it to operate independently as a basis for deciding comparative punishments. Any increase or decrease in the severity-ranking of a penalty on the scale alters how much censure is expressed—and hence needs to be justified by reference to the seriousness of the criminal conduct involved.

4. THE CRITERIA FOR PROPORTIONALITY

When we say sanctions should be 'proportionate', what does that mean? Is there any particular quantum of punishment that is the deserved penalty for

---

is required for liability—because it is thought preferable that the actor bear the burden of loss, rather than the person he has injured through his carelessness. But it is not part of the point—the very significance—of an ordinary civil recovery to convey disapprobation. Indeed, the fault requirement in civil recoveries could be eliminated, were there a practicable alternative method for distributing the burden of loss: one such method is the no-fault scheme, where those who suffer harm are compensated through State insurance.

crimes of a given degree of seriousness? If not, what guidance does the principle give?

To answer such questions, let me advert to the distinction between ordinal and cardinal proportionality.* *Ordinal proportionality* relates to comparative punishments, and its requirements are reasonably specific. Persons convicted of crimes of like gravity should receive punishments of like severity. Persons convicted of crimes of differing gravity should receive punishments correspondingly graded in their degree of severity. These requirements of ordinal proportionality are not mere limits, and they are infringed when persons found guilty of equally reprehensible conduct receive unequal sanctions on ulterior (e.g. crime prevention) grounds. The ordinal proportionality requirements are readily explained on the reprobative conception of punishment just set forth. Since punishing one crime more severely than another expresses greater disapprobation of the former crime, it is justified only to the extent the former is more serious.

Ordinal proportionality involves three sub-requirements, which are worth summarizing briefly.** The first is *parity*: when offenders have been convicted of crimes of similar seriousness they deserve penalties of comparable severity. This requirement does not necessarily call for the same penalty for all acts within a statutory crime category—as significant variations may occur within that category in the conduct's harmfulness or culpability. But it requires that once such within-category variations in crime-seriousness are controlled for, the resulting penalties should be of the same (or substantially the same) degree of onerousness.*** (This parity requirement has one possible exception, concerning the role of prior convictions, that will be touched upon later.****)

A second sub-requirement is *rank-ordering*. Punishing crime Y more than crime X expresses more disapproval for crime Y, which is warranted only if it is more serious. Punishments should thus be ordered on the penalty scale so that their relative severity reflects the seriousness-ranking of the crimes involved.

The third sub-requirement concerns *spacing* of penalties. Suppose crimes X, Y, and Z are of ascending order of seriousness; but that Y is considerably more serious than X but only slightly less so than Z. Then, to reflect the conduct's gravity, there should be a larger space between the penalties for X and Y than for Y and Z. Spacing, however, depends on how precisely comparative gravity can be calibrated—and seriousness-gradations are (as we will see in Chapter 4) likely to be matters of rather inexact judgement.

Scaling penalties calls also for a starting point. If one has decided what the

---

* That distinction is outlined in A. von Hirsch, *Past or Future Crimes* (1985), ch. 4.

** For a fuller discussion of these requirements, see ibid., chs. 4–7; and von Hirsch, 'Proportionality in the Philosophy of Punishment' (1992).

*** For fuller discussion, see the end of Ch. 8.

**** See Ch. 7, below.

penalty should be for certain crimes, then it is possible to fix the sanction for a given crime, X, by comparing its seriousness with the seriousness of those other crimes. But no quantum of punishment suggests itself as the uniquely appropriate penalty for the crime or crimes with which the scale begins. Why not? Our censure-oriented account again provides the explanation. The amount of disapproval conveyed by penal sanctions is a convention. When a penalty scale has been devised to reflect the comparative gravity of crimes, altering the scale's magnitude by making *pro rata* increases or decreases represents just a change in that convention.

Not all conventions, however, are equally acceptable. There may be limits on the severity of sanction through which a given amount of disapproval may be expressed, and these constitute the limits of *cardinal* or non-relative proportionality. Consider a scale in which penalties are graded to reflect the comparative seriousness of crimes, but in which overall penalty levels have been so much inflated that even the lowest-ranking crimes are visited with prison terms. Such a scale would embody a convention in which even a modest disapproval appropriate to low-ranking crimes is expressed through drastic intrusions on offenders' liberties. If suitable reasons can be established for objecting to this convention (for example, on grounds that it depreciates the importance of the rights of those convicted of such low-ranking crimes*), a cardinal—that is, non-relative—constraint is established.

The cardinal–ordinal distinction explains why one cannot identify a unique 'proportionate' sanction for a given offence. Whether x months, y months, or somewhere in between is the appropriate penalty for (say) armed robbery depends on how the scale has been anchored and what punishments have been prescribed for other crimes. The distinction explains, however, why proportionality becomes a significant constraint on the ordering of penalties. Once the anchoring points and magnitude of the penalty scale have been fixed, ordinal proportionality will require penalties to be graded and spaced according to their relative seriousness, and require comparably-severe sanctions for equally reprehensible acts.

---

* For fuller discussion, see Ch. 5.

3
'Dominion' and Censure
Co-author: *Andrew Ashworth*

AMONG recent critiques of the desert model★ is one by a sociologist and a philosopher, John Braithwaite and Phillip Pettit.★★ The book's authors consider the whole idea of sentences apportioned to the gravity of offences to be mistaken, and offer an alternative, consequentialist theory of justice that supposedly would decide sentencing policy better. The volume interests us for two reasons. First, its authors' own theory differs from traditional penal utilitarianism in that it tries (albeit otherwise than we would) to emphasize notions of autonomy and choice. Second, the authors favour a blaming role for the criminal sanction—but one that would point away from, not toward, sanctions that are proportionate.

1. THE THEORY IN BRIEF

The traditional utilitarian calculus purports to decide punishments by weighing harms in the aggregate: the injury caused by crime (and fear of crime) is to be 'balanced' against the pains suffered by those punished (and the financial and social costs of law enforcement). The drawbacks of this approach are familiar enough. Aside from the manifold problems of assessing and comparing the harms involved, such a scheme can support drastic interventions against the few, if the net benefits to the many are large enough. Draconian punishments become permissible, provided only that they are 'optimizing'.[1]

Braithwaite and Pettit wish to retain the forward-looking and aggregative

---

★ Other critiques include Nicola Lacey's *State Punishment* (1988), and Nigel Walker's *Why Punish?* (1991). Lacey's criticisms, made from a communitarian perspective, are examined in our *Principled Sentencing* (1992), ch. 7. Walker adopts a traditional utilitarian perspective.

Previously, the main theoretical challenge to desert theory has come from Norval Morris, who suggests that proportionality sets only broad outer bounds on punishments, within which sentences would be decided on other grounds. See his *Madness and the Criminal Law* (1982), ch. 5. For a reply to his views, see Ch. 6 below, and more fully, A. von Hirsch, *Past or Future Crimes* (1985), chs. 4, 12.

★★ *Not Just Deserts: A Republican Theory of Punishment* (1990).

features of the utilitarian calculus. They are troubled, however, by the calculus's seeming disregard of the person. Their solution, essentially, is to retain the calculus but change its measure from utility to something that would give greater emphasis to persons' capacity for choice. They term this something else 'dominion'. That notion is not defined clearly, but seems to be a kind of satisfaction that stresses agency and republican virtue: that the actor be able to determine his own course of living, within social and political institutions which he or she, as a citizen, participates in shaping. A harsh system of punishments might be efficient in preventing harm, they argue, but it is not likely to further dominion: fear-provoking sanctions would diminish, not enhance, citizens' sense of control over their own lives.[2]

With this modification, the calculus would be retained. One would estimate the loss of 'dominion' (that is, self-determination) caused by crimes and the fear they inspire, and also try to gauge how much punishments diminish the dominion of those who are or may be punished. Optimum penalties would be those which yield the least feasible loss of dominion.[3]

Translating such a conception into a scale of sanctions presents manifest difficulties. Gauging gains and losses of 'dominion' would be still more elusive than totting up costs and benefits in the usual utilitarian fashion. What guidance, then, can Braithwaite and Pettit provide for fixing punishments?

Proportionality, the authors suggest, may largely be disregarded: penalties need not be graded according to crimes' gravity. This follows from the forward-looking character of the suggested rationale. Offenders who are deemed dangerous, for example, may receive substantially longer terms than non-dangerous offenders convicted of comparably-serious crimes— since their extra confinement will protect the dominion of potential victims.[4] The gravity of the offence should matter only in the setting of some upper limit on the punishment's severity, beyond which such extra terms could not extend.[5]

In the preceding chapter, the argument for proportionate sanctions was made on reprobative grounds: punishments convey censure or blame, and hence should be ordered according to the degree of blameworthiness of the conduct. Braithwaite and Pettit speak approvingly of reprobation; indeed, Braithwaite, in a separate volume,[6] offers a whole theory of 'shaming' responses to crime. However, the two authors hold that shaming can largely be uncoupled from the severity of sanctions. It would be possible, they assert, to have substantial inequalities in the punishment of persons whose crimes are comparably serious, and still express reprobation by other means.[7]

If not desert, what else should determine penalties? The authors propose a 'decremental strategy', according to which existing penalty-levels would gradually be scaled down, until the point is reached where those reductions produce measurable increases in crime. Such penalty reductions would enhance aggregate 'dominion', because convicted offenders would suffer less intrusion on their liberties without any increase in victimization of other citizens. However, they say, 'when the first evidence is produced which justifies the belief that the accumulation of criminal justice cuts has lifted crime rates, further cutting should be stopped'.[8]

The scheme also expressly allows political considerations to affect the setting of penalties, as the following passage indicates:

A decrease in average sentences may not produce an increase in crime, but under the sway of law-and-order rhetoric the community may become convinced it has dramatically increased crime. An epidemic of fear of crime in the community might grow to be a serious threat to the dominion of citizens who are suddenly afraid to walk in their neighborhood at night. Vigilantism might arise. Such consequences cannot be predicted in advance; but neither can they be ignored. A decrementalist strategy with constant monitoring of system outputs will move slowly, watching for any such community reactions and attempting to cope with them.[9]

This passage seems to be saying, not merely that politics will in fact influence sentencing policies, but that community fears (even if unwarranted) *should* affect punishment levels—because such fears affect citizens' sense of control over their own lives. Such a view is consistent with the authors' wholehearted consequentialism, but it is troublesome nevertheless (as we shall see shortly).

2. 'DOMINION' AND AGGREGATION

As is evident from this sketch, the authors wish to improve upon the utilitarian calculus by replacing its units of utility with gains and losses in 'dominion'. The theory, they assert, will thus give greater protection to individuals.

Are they correct? There are contexts, outside the criminal law, where shifting the calculus from satisfaction to 'dominion' might help. One such context involves paternalistic interventions—restricting the liberty of persons for their own supposed good. Conventional utilitarianism is capable of supporting sweeping paternalist powers: it need merely be argued that the persons involved gain more from being prevented from harming themselves than they would by having their choices respected.[10] A shift to 'dominion' would rule this out: if choice is the measure, it is difficult to argue that an ambitious paternalism promotes the self-determination of those affected.

Punishment is different, however. It is true that punishment interferes with the choices of the punished. To the extent of its preventive efficacy, however, it also fosters other persons' choices—by safeguarding them from unwanted victimization. If more people are safeguarded than intruded upon, why not proceed? The forward-looking and aggregative features of the authors' theory thus give licence to punish to the full extent which would cause potential victims' net gain in 'dominion' to exceed the losses in dominion of those punished.

What, however, of the authors' suggested maximum limits on permissible punishment? They offer no persuasive reason for those limits' existence. Braithwaite and Pettit cannot argue, as we do, that disproportionately severe penalties visit undeserved blame, for they wish to detach reprobation from the quantum of sentence. They suggest, instead, that upper limits are needed to give citizens a sense of security against the punishment system.[11] However, eliminating such limits—or making them easily permeable when dealing with dangerous offenders—could arguably enhance potential victims' sense of security against predatory conduct.

The aggregative character of the authors' 'dominion' theory also jeopardizes their proposals for reducing overall punishment levels. They urge, as we saw, a 'decremental strategy', according to which punishments are to be reduced across the board, until the point is reached where crime rates are thereby caused to rise.[12]* But their penal theory does not necessarily support such decreases. The 'decremental' strategy could easily be stood on its head: crime rates are at unacceptable levels now, arguably, so why not *lift* sanctions incrementally until the incidence of crime begins to fall? This latter possibility is left open all the more by the fact that the authors permit consideration not only of crime rates but also fear of crime. Might someone not argue, with Ralf Dahrendorf,[13] that increased sanction levels would reassure the public and reduce fear, even if they did not actually affect crime rates? It is, in short, far from clear whether promotion of 'dominion' in the aggregate would best be achieved by low punishment levels but possibly a fairly high incidence of crime or fear of crime; or by higher punishment levels, were those able to reassure citizens better.

Desert theory does permit a 'decremental' strategy in the setting of the anchoring points of a penalty scale[14]—and a version of such a strategy will be proposed in Chapter 5. However, penalties would have to be ordered according to the gravity of the criminal conduct, and 'optimization' is not presupposed. It does not have to be assumed that the penalty levels that are adopted should lead to a total increase in satisfaction (or 'dominion' or whatever else) in the world.**

---

* The strategy assumes a considerable capability to measuring deterrent effects, despite the authors' elsewhere-stated cautions about how little we know about deterrence. When penalties are reduced and crime rates begin to rise, how does one tell if the rise is caused by the penalty reductions or by other factors? The authors reply that 'criminology should be able to cope with these policy-monitoring challenges in the same admittedly rough ways that policy-makers cope [with other kinds of assessments].' But this is surely optimistic, considering the extensive criminological literature on the difficulty of measuring deterrent effects. In the absence of an adequate assessment capability, the strategy is apt to boil down to one of lowering penalties until crime rates rise (for whatever reason). That, unhappily, will limit the scope of penalty reductions.

** For further discussion of this 'optimization' question, see Ch. 5, below.

3. DETACHING CENSURE FROM THE SEVERITY OF SANCTIONS

Let us turn, then, from the authors' general conception of punishment to their particular views on censure. These are of interest to us, because the link between blaming and proportionality is questioned. Punishment, it was suggested in the preceding chapter, expresses censure; and the severity of the punishment conveys how much the conduct is disapproved. It is because of this expressive feature, that the quantum of sentence should reflect the degree of blameworthiness of the criminal conduct. Braithwaite and Pettit challenge this reasoning: while favouring the idea of blaming offenders, they seek to separate blame from the quantum of sentence.

The authors take an instrumental view of censure—akin to the notion of 'denunciation' in the traditional English legal literature.[15] Censure or denunciation is seen as a kind of stimulus to law-abidingness, that operates because people tend to be more reluctant to engage in conduct when it is condemned.[16] On this view, it scarcely matters how the disapproval is expressed—whether through the severity of the punishment, or through other means such as adverse publicity conveyed during trial, or penalties that are highly visible.

Moreover, the authors largely identify censure with something else: stigma. What counts is not how much disapproval the State is expressing through punishment, but how that disapproval 'takes': how much social obloquy actually attaches to the particular defendant.[17]

It is these two assumptions—the instrumentalist view of censure and the fusion of censure with stigma—that permit the authors to uncouple 'censure' from the quantum of sentence. Suppose crime X, a street crime, and crime Y, a white-collar crime, are of comparable seriousness; but that white-collar criminals tend to be more respectable (and hence more sensitive to bad publicity) than street criminals. Then, according to the authors, there is nothing wrong with punishing the white-collar criminals more leniently but publicizing their crimes,—as that will produce a roughly equivalent amount of stigma. Moreover, it is not even required that there be parity in the degree of stigmatization. Under Braithwaite's and Pettit's view, censure has mainly the function of strengthening people's inhibitions against crime, so that inequality in stigmatization among comparable cases does not matter much unless it interferes with this function.

This entire account is mistaken, in our judgement. It misconceives both the logic of censure, and the connection between censure and the criminal sanction. Censure, in the sense of moral disapprobation, is desert-oriented by nature. One is entitled to condemn only if one has reason to believe that the conduct is wrong, and only to the degree of its wrongfulness. The person censured has reason to object, if the conduct is not wrongful, or if the degree of the condemnation is excessive given how reprehensible the conduct is. The conscientious condemnor should consider such protests, and alter his response if he judges they are correct.[18]

It is this desert-oriented character that gives blaming its distinctive force. Why should I be concerned when another censures me for what I do? Why might I feel ashamed, or make extra efforts at self-control? Not, surely, just to please the person. It has something, instead, to do with the fact that I assume the disapprobation to be an authentic expression of the condemnor's ethical judgements, and that I have a certain regard for his or her capacity to make such judgements. Were I to think that the censure was merely an attempt in moralistic garb to manipulate my behaviour, I should surely be less impressed.

Braithwaite's and Pettit's purely instrumental notion of blaming—that one blames only to influence others and then only to the extent needed for such influence—would undercut this logic. If one suits the degree and manner of blaming to whatever works, one is treating the condemnation as just another inducement to comply, and not as an appeal to the actor's capacity for moral assessment of his or her own conduct. Since the degree of condemnation is no longer suited to the gravity of the act, there is no reason (other than mere fear of other persons' unfavourable attitudes) why a conscientious actor ought to share the judgement.

In punishment, one is speaking of a State-inflicted response, but the logic still holds. The decision to visit conduct with censure needs to be supported by a judgement on the merits that the conduct is reprehensible, and is so to the degree to which it is visited with blame. If A and B commit criminal conduct of an equal degree of reprehensibleness, it is still morally problematic to visit A with more censure than B—even if doing so would have the desired degree of influence over their conduct.

This brings us to the link between censure and the severity of punishment. Punishment, by its very logic, expresses disapproval, and the quantum of the penalty reflects how much disapproval is expressed. Giving offender A a severer sentence than offender B necessarily involves the State's blaming B more. Braithwaite and Pettit deny this link, and assert that 'denunciation' can operate independently of the severity of the penalty: in their words:

Denunciation, we believe, is determined less by the length of sentence than by whether the trial is reported by the media, whether it is held in open court in the presence of significant others, how many of the offender's acquaintances come to know of the conviction throughout the rest of his life.[19]

This claim, however, rests on the authors' conflation of censure with stigma. Were one speaking of the defendant's reputation, it is conceivable (though seldom likely) that the other factors they mention would matter more than the severity of the sentence itself. But censure concerns how much disapprobation is expressed through the sanction, and that depends on its severity. If the judge gives A a long prison sentence and B a modest fine, A is thereby visited with more censure than B. It does not matter that the judge smiles sweetly when giving A his prison sentence, whereas he gives B a furious lecture from the bench (that attracts media attention) when he imposes the fine.

The authors assert that if the severity of the sanction matters at all, then it is the *maximum* sentence that should reflect the blameworthiness of the conduct. Thus:

The state can effect the desired moral education by providing for high maximum penalties . . . to signify the seriousness of [the more grave kinds of] crime. At the other extreme, when the state does not provide for severe penalties . . . the community is apt to get the message that this kind of crime is not regarded very seriously.[20]

The authors are misled here by their purely functional view of censure, for the issue is not what would best prompt the public to greater law-abidingness. The sentence visits a certain amount of disapproval on its recipient, depending on its actual severity. When A is punished more than B, it signifies that his conduct is worse,—and it does not matter that both offenders *might* have been given the same maximum penalty.

Could censure ever operate independently from the severity of the sanction? It is possible to imagine scenarios in which it might, but these concern institutions quite different from *punishment*. Were the incidence of predatory conduct low enough, for example, the society might not require a State-inflicted penal sanction at all. Various informal, non-criminal sanctions might suffice to keep the conduct at tolerable levels. Any formal response might be purely symbolic—an expression of blame not tied to the visitation of hard treatment.* This scenario, however, would not be helpful to Braithwaite and Pettit. There would be no criminal sanctions to allocate at all—even under their own 'republican' theory.**

Alternatively, hard treatment could be retained, but kept separate from reprobation. There would be a symbolic censuring response—followed by a *separate* non-condemnatory sanction. That scheme, however, would have its own manifest difficulties. We ourselves would object to it for reasons already touched upon earlier: the non-condemnatory sanction would be pure 'tiger control' which does not address the actor as a moral agent.[21] Braithwaite and Pettit might not worry about this, but there are practical difficulties as well. The initial, symbolic response could easily deteriorate into an insignificant formality, as the parties and the public await the second, non-condemnatory imposition, with its more dramatic and visible consequences. (This is especially likely in modern societies lacking a high

---

* See Ch. 2, above.
** Moreover, the formal expressions of reprobation should still, under this scenario, reflect the gravity of the conduct. See Ch. 2.

degree of cohesion, where purely symbolic messages from any agency, including the State, do not carry much credibility.) Such a two-step scheme could thus be workable only in a quite different social environment, and, in any case, it is not the criminal sanction as we understand it. What is unsustainable is the authors' claim: that we can keep punishment (with its closely-linked blaming implications), and *still* justly separate the blame from the severity of the sanction.

4. UTOPIANISM AND COMPREHENSIVENESS

Braithwaite's and Pettit's theory also disturbingly mixes utopianism with the here-and-now. Their proposals for progressive penalty reductions, as they themselves admit, are ambitious and not easily achievable, given the politics of most jurisdictions.[22] Nevertheless, they urge that desert constraints should be jettisoned immediately, even in places unpropitious for their brand of decrementalism. In today's sentencing systems, however, abandoning desert constraints could open the way to consequentialism of the most frightening sort. It would not be mild penalties, but quite substantial ones that could be allocated without regard to the degree of blameworthiness of the conduct.

A final point about scope: desert theory addresses sentencing, not the criminal justice system in general. We doubt, however, whether this is a weakness. Different stages of the criminal process raise normative questions that are different. Contrast sentencing with pre-trial detention. A sentence can be fixed proportionately to the blameworthiness of the criminal conduct. Prior to trial, however, it has not been determined that the actor has done anything blameworthy at all, and certainly there has been no occasion yet for State censure to attach. We would have a difficult time developing a single rationale that applies equally well to sentencing and pre-trial confinement.

Braithwaite and Pettit say their own theory has the virtue of comprehensiveness: it holds for all stages of the criminal process. Indeed, one might go further and try to apply their 'dominion' calculus to all social-policy decisions, as traditional utilitarians did with Bentham's principle of utility. Such sweepingness has its costs, however. It makes the rationale indeterminate: almost any sentencing policy (or for that matter, police or prosecution policy) might be defended on the basis that it promotes 'dominion'. It also causes one to overlook important moral distinctions that would become apparent from a closer look at a particular kind of institution—for example, at the blaming features of the sentence.

The authors' demand for comprehensiveness seems oddly out of date. Philosophers have become increasingly sceptical of sweeping, foundationalist moral and social theories.[23] The conception that purports to answer every question is apt to yield answers that are meagre at best and at worst, plain wrong. It is the richness of particular social institutions that provide the necessary clues for constructing normative theories. There is no patentable formula for dealing with criminal policy, or with other hard social questions, in general.

4

Seriousness and Severity

THE principle of proportionality requires the *severity* of penalties to be determined by reference to the *seriousness* of crimes. In order to apply the principle, we need to be able to gauge how serious various crimes are, and how severe are various sanctions. Where does burglary rank on a scale of crimes' gravity? Is a given penalty—say, home detention for so many weeks—a severe sanction?

While I have addressed crime-seriousness before,[1] my views on its harm component have since altered (as will be explained shortly). How to assess sanctions' severity has scarcely been dealt with by desert theorists. This neglect is not surprising, perhaps, because attention was focused so much on the use and limits of imprisonment, and that sanction's onerousness appears measurable in large part by its duration. Now, however, there is increasing interest in non–custodial penalties, and these are more hetero-geneous in character. How, for example, can a given number of days of community service be compared in severity with a fine of that number of days' earnings?

1. GAUGING CRIMES' SERIOUSNESS

Ordinary people, various opinion surveys have suggested, seem capable of reaching a degree of agreement on the comparative seriousness of crimes.[2] And rule-making bodies which have tried to rank crimes in gravity have not run into insuperable practical difficulties. Several US state sentencing commissions (those of Minnesota, Washington State, and Oregon) were able to rank the seriousness of offences for use in their numerical guidelines.[3] While the grading task proved time-consuming, it did not generate much dissension within the commissions.

Less satisfactory, however, has been the state of the theory. What criteria should be used for gauging crimes' gravity? The gravity of a crime depends upon the degree of harmfulness of the conduct, and the extent of the actor's culpability.[4] Culpability can be gauged with the aid of clues from the substantive criminal law. The substantive law already distinguishes intentional (i.e. purposive, knowing, or reckless) conduct from negligent. It should not be too difficult in principle to develop, for sentencing doctrine, more refined distinctions concerning the degree of purposefulness, indifference to consequences, or carelessness involved in a sentenced actor's conduct.[5] The doctrines of excuse in the substantive criminal law could also be drawn upon to develop theories of partial excuse—for example, of partial duress and diminished capacity.[6]

The harm dimension of seriousness is more puzzling, however, as the substantive law provides no assistance: it does not formally distinguish degrees of harm. How, then, can one compare the harmfulness of acts that invade different interests, that is, compare crime X that invades property interests and crime Y that chiefly affects privacy? In a 1985 work,[7] I suggested (drawing on ideas of Joel Feinberg[8]) that harms should be compared according to the degree to which they characteristically intrude upon people's *choice*. Violence and graver kinds of victimizing fraud, for example, would qualify as particularly harmful because persons who suffer serious bodily injury or are rendered destitute have their options so drastically curtailed.

I have come recently to feel, however, that a choice-based standard is somewhat artificial. Consider the interest (which we doubtless feel is very important) in avoiding intense and protracted physical pain. Such pain would interfere with choice—with the breadth of options a person might have. But that scarcely accounts for why we think this interest so important. It would be more natural, instead, to judge this interest in terms of how it affects the quality of a person's life. What really is wrong with extended physical suffering is that it makes for an awful life.

Nils Jareborg and I have, in a 1991 essay,[9] sketched a theory of harm that focuses on such questions of quality of life. According to our theory, victimizing harms are to be ranked in gravity according to how much they typically would reduce a person's *standard of living*. We use that term in the broad sense suggested by Amartya Sen, which reflects both economic and non-economic interests.[10]

The living standard is one of a family of related notions, including well-being, that refer to the quality of persons' lives. Well-being, however, can be highly personalized: my well-being depends on my particular focal aims, and to the person who wants to devote his life to contemplation and prayer, material comfort and ordinary social amenities may matter little. However, the living standard, in Sen's sense, does not focus on actual life-quality or goal achievement, but on the *means or capabilities* for achieving a certain quality of life. Some of these means are material (shelter and financial resources) but others are not (good health, privacy, and the like). It is also standardized, referring to the means and capabilities that would *ordinarily* promote a good life. Someone has a good standard of living if he has the health, resources, and other means that people ordinarily can use to live well.

Using the living-standard as a way of gauging criminal harms has a number of advantages. First, it appears to fit better the way we usually judge such harms. Why is mayhem more harmful than burglary? Not just because the maimed person's options have been narrowed more. It is because the overall quality of the person's life has been more adversely affected. Second, the living-standard idea permits drawing from a richer array of experience, including experience outside of the criminal law. If a typical person's standard of living is at issue, the interests affected by criminal acts can be compared with those affected by non-criminal occurrences: we can ask how the harmfulness of an arson compares with that of an accidental fire. Finally, a living-standard analysis would allow for cultural variation. Different social living-arrangements can affect the consequences of a criminal act, and normative differences among cultures can affect the impact of those consequences on the quality of persons' lives. The harmfulness of burglary, for example, depends on the degree to which the home is ordinarily the focal point for people's private existences. A living-standard analysis thus could, in another society, lead to a different rating for burglary—if the home has a different social role and if another valuation is given to privacy.

The mechanics of a living-standard analysis of harms are described in Jareborg's and my essay,[11] and need only be summarized here. The suggested technique involves parcelling out the various kinds of interests that offences typically involve. After the interests involved in a given type of offence are identified, their importance is judged by assessing their normal significance for a person's living standard.*

Most victimizing offences involve one or more of the following interest-dimensions: (1) physical integrity, (2) material support and amenity, (3) freedom from humiliation, and (4) privacy. A simple residential burglary, for example, chiefly involves material amenity and privacy. The material loss would consist of what is stolen, plus the inconvenience and expense of repairs. The privacy-loss consists of the intrusion of a stranger into the person's living-space. To rate the harmfulness of the conduct, the living-standard criterion would be applied to each dimension, successively. In the case of the burglary, the analysis would thus begin with its material-amenity dimension. Here, the impact on the living standard would be rather small: not much is taken in the typical burglary, so that the person's material well-being is scarcely affected. Next, the privacy dimension would be considered. Here, the rating might well be somewhat higher, given the importance of privacy to a good existence, and the extent to which an intrusion into the home affects privacy. The attraction of this mode of analysis is that each dimension involved—physical integrity, material

---

* Our analysis—which addresses both actual and risked harms—is designed to apply to crimes that typically involve identifiable victims, as burglary and robbery do. However, we suggest in our essay some ways it might be extended to crimes of other sorts—for example, those that risk injury to unidentified persons, or affect 'collective' interests.

support, privacy, or whatever—can ultimately be assessed in terms of a common criterion: that of impact on the living standard. This means that, in the burglary, one can compare the living-standard impact of the material loss (rather minor) with that of the privacy–intrusion (arguably, somewhat greater). A burglary can also be compared with another victimizing offence involving different interests: say, an assault, where physical integrity and freedom from humiliation are primarily involved.

To aid in this analysis, we also suggest grading the living standard itself. We would use four living-standard levels: (1) subsistence, (2) minimal well-being, (3) 'adequate' well-being, and (4) 'enhanced' well-being. The first, subsistence, refers to survival but with maintenance of no more than elementary human capacities to function—in other words, barely getting by. The remaining three levels refer to various degrees of life quality above that of mere subsistence. The function of the four gradations is to provide a rough measure of the extent to which a typical criminal act intrudes upon the living standard. To take an obvious example, aggravated assault threatens subsistence, and thus is substantially more harmful than theft of a kind that still leaves the person with an 'adequate' level of comfort and dignity.

This analysis is to be used chiefly in gauging the standard harm involved in various categories or sub-categories of crimes. The point is to assess the injuriousness of typical instances of (say) residential burglary, or residential burglary of a certain kind (one might, say, distinguish simple burglary from ransacking). The living-standard relates, as we noted, to the *standardized* means or capabilities for a good life—not to the life-quality of particular persons. Deviations from such standardized assessments could be made in special circumstances, but only where the differences from the ordinary case are fairly marked.★

How helpful this way of gauging harms is can only be ascertained by trying it. Suppose, for example, that a deliberative body is writing guidelines, of either an advisory or more binding character, for sentencing decisions. One of its tasks would be to rate the gravity of offences. To date, when sentencing commissions in the USA undertook this rating task, they had to rely largely on the members' intuitions, for not much in the way of reasons for the ratings could be provided.[12] The living-standard analysis could help supply reasons. After the members of the guideline-writing body rank offences intuitively, according to their sense of the crimes' seriousness, they might—with respect to each offence (or each victimizing offence)—try to identify the interest-dimensions involved, and then rate those interests' importance to the living-standard. This would provide some formal assessment of the harm involved in the offences. The commission should then make explicit judgements about the degrees of

---

★ For further discussion of this issue, see Ch. 8.

culpability involved. Having undertaken this formal analysis, the commission could examine to what extent its resulting seriousness-ratings differed from the members' original (intuitive) ratings.[13] Such a procedure provides only a guide to judgement, not a formula. In applying the living-standard analysis, the members of the commission still would have to rely on their best judgement of how different kinds of conduct might typically affect the living standard; only now, that judgement would be guided by explicit principles.

2. GAUGING PUNISHMENTS' SEVERITY

Grading sanctions presupposes an ability to judge their comparative severity. While prison sanctions can be compared by their duration, non-custodial sanctions' onerousness depends to a greater degree on their intensity. Three days of community service may be tougher than three days' probation, but not so tough as three days of home detention.

A number of studies have attempted to measure sanction severity through opinion surveys.[14] A selected group of respondents is shown a list of penalties of various sorts, and asked to rate their severity on a numerical rating scale. The surveys tend to show a degree of consensus. These ratings, however, do not attempt to elucidate what is meant by severity; to elicit respondents' reasons for their rankings; or to assess the plausibility of those reasons. It is necessary to consider what *should* be the basis of comparing penalties, i.e. develop a theory for gauging severity.

A possible account of severity would be one that depends on sanctions' degree of unpleasantness. On this view, surveys could be employed that simply ask people how unpleasant they think various penalties are. Unpleasantness or discomfort, arguably, is ultimately subjective: a matter of how deprivations typically are or might be experienced. Castor oil is nasty stuff for no better reason than that most people find it tastes nasty. If penalty X is generally perceived to be more onerous than penalty Y, that would make it so.

The castor oil analogy, however, is misleading. What makes punishments more or less onerous is not any identifiable sensation; rather, it is the degree to which those sanctions interfere with interests that people value. The unpleasantness of intensive probation supervision, for example, depends—not on its 'feeling bad' in some immediate sense—but on its interfering with such interests as being in charge of one's own life or moving about as one chooses.

In matters that are purely subjective such as tastes or smells, one cannot challenge someone's judgement or demand reasons. If (as I am told) the ordinary Englishman hates peanut butter, I can suggest that it tastes good on toast when eaten for breakfast. But if my English interlocutor tells me that he and his friends cannot stand the sticky stuff, that is the end of it. And if the English dislike of peanut butter seems to fit incongruously with a liking for other odd breakfast foods such as kippers or Marmite, so be it.

Judgements concerning punishments' severity have, I think, a different logic. If I describe intensive probation supervision, and my interlocutor says it sounds pretty soft, that is not the end of the discussion. I can point out that intensive supervision considerably limits a person's autonomy and privacy, and that those are things that ordinarily matter to a tolerable life. He or she might disagree in varous ways: for example, by suggesting that I overestimate the degree to which the supervision, as it in fact operates, does intrude upon the person's choices. But the reply of the Englishman who hates peanut butter—'well, that's just the way I (or we) feel'—is no answer.

Is it not true, however, that punishments need to be subjectively unpleasant? In Chapter 2, I offered a rationale of punishment that appealed to two functions: its capacity to convey censure and to its effect as a practical disincentive. Do not both functions presuppose sanctions which people dislike? Without punishment's being unpleasant, they would not be able to claim the attention of the person whose conduct is being censured, nor the act as a disincentive. This means penal measures that feel unpleasant would be preferable to hypothetical penalties that have a comparable impact on people's lives but (for whatever reason) would not be experienced as disagreeable.* That, however, does not mean the *measure* of severity must be a subjective one. It should be recalled that my penal theory is not one according to which punishments must be graded to maximize their potential deterrent effects. Its criterion for severity thus need not track the subjective unpleasantness of penalties exactly.

An alternative (and in my judgement preferable)[15] approach would be to rely on an interests-analysis, comparable to the one just suggested for gauging crime-seriousness. The more important the interests intruded upon by a penalty are, on this theory, the severer it is. Penalties could be ranked according to the degree to which they typically affect the punished person's freedom of movement, earning ability, and so forth. The importance of those interests could then be gauged according to how they typically impinge on a person's 'living standard'—in the sense of that term sketched earlier in this chapter. Such an interests-analysis seems to fit the way we often discuss punishments' severity. To explain why long-term imprisonment is a severe sanction, for example, it is natural to point out how much its deprivations typically impinge on the quality of a person's life.

To apply the living-standard idea to penalties, there would have to be modifications in the analysis. When rating crimes, the main interests are (as noted earlier) physical integrity, material amenity, and so forth. When

---

* One could imagine such penalties. Suppose a latter-day Dr Gillotin invents a penalty akin to cigarette-smoking: it is not unpleasant to undergo at the time, but reduces one's life-expectancy. On a living-standard theory this sanction would be severe, as it affects survival itself. (Happily, such a sanction does not exist.)

rating punishments, those interests often are different: for example, the interest in freedom of movement becomes more salient in sanctions such as incarceration, home detention, and intensive probation supervision. A new taxonomy of interests would thus have to be developed, and the idea of the living standard then applied to rating the importance of those interests.*

Adopting an interest-analysis approach means that the assessment of severity is not made dependent on the preferences of particular individuals. The living-standard, as noted earlier, refers to the means and capabilities that *ordinarily* assist persons in achieving a good life. If a given interest is important in this sense to a decent existence, it would warrant a high rating—notwithstanding that some persons might choose to go without it. Imprisonment qualifies as a severe penalty—because the interests in freedom of movement and privacy it takes away are normally so vital to a good life—despite the fact that a few defendants might happen to be claustrophiliacs.

This account thus answers an objection that Nigel Walker has raised to desert theory: that one cannot grade penalties' onerousness, because individuals' subjective perceptions of painfulness vary. 'If the penalty is imprisonment', he asks, 'how much does loss of freedom mean [to the particular defendant]?'[16] Severity, however, is simply not a matter of (variable) subjective unpleasantness. If the freedom of movement that incarceration interferes with is an important interest—in the sense of its importance as a standardized means to achieving a certain living-standard—then its deprivation is severe even if a particular person might have different personal sensitivities.

---

* How do these conclusions affect the usefulness of opinion surveys, as a means for gauging severity? On the subjectivist view I have rejected (namely, that severity is a matter of penalties' unpleasantness), surveys would obviously have been useful, as the way of gauging how unpleasant various sanctions are felt to be. On my living-standard view, opinion surveys would not have a comparable dispositive effect. What matters is how much a given penalty *does* affect the ordinary person's living standard. This is a matter, not of surveying what people think, but of analysing the effects of the penalty on the quality of persons' lives.

Surveys might be more interesting, however, were a living-standard analysis built into them. Instead of merely being asked to rate sanctions' severity, respondents could be asked specific questions relating to what interests penalties intrude upon, how those intrusions would affect the quality of life, and why so. Research of this kind has not yet been attempted, however.

5
Anchoring the Penalty Scale

How should the penalty scale be anchored? The question has daunted desert theorists, and little has been written on it. Yet its importance is obvious. Even when penalties have been arranged on the scale according to the comparative gravity of offences, the scale's magnitude still must be decided. What should be the scale's severest and most lenient penalties? To what extent may penalties be deflated pro rata? What considerations militate against rateable increases? Without answers to such questions, it is not clear whether a proportionate scale should involve reducing penalties overall, toughening them, or keeping sanction levels where they are today.

In dealing with anchoring a scale, two main issues arise, which I shall consider successively in this chapter. The first concerns how the principle of proportionality constrains the scale's magnitude. The second relates to the anchoring of the penalty scale *within* applicable cardinal-proportionality constraints.

1. CARDINAL-PROPORTIONALITY CONSTRAINTS

The idea of cardinal proportionality has only been touched upon thus far.★ Exactly *why* might a penalty scale as a whole be deemed disproportionately severe or lenient? And how do those reasons relate to our guiding idea of penal censure?

The Upper Bounds of Cardinality

Let me consider, first, the upper bounds of cardinal proportionality—that is, the constraints against inflating penalties pro rata above a certain level. Suppose a hypothetical penologist, Dr Draco, proposes the following penalty scale. Crimes are ranked in seriousness from, say, '1' (the least serious) to '20' (the most). Ordinal proportionality is scrupulously observed: all penalties are ranked and spaced consistently with the gravity of the offences involved. However, Draco's scale is harsh. The least serious

★ See the end of Ch. 2, above.

crimes, with a rating of only '1', carry a penalty of several months' imprisonment; intermediate-level crimes are punishable by several years in confinement; and serious crimes carry several decades. Such a scale would seem to be disproportionately severe. But why so? The answer involves three steps.

First, low-ranking crimes (that is, those that are rated low on the seriousness scale) are not very reprehensible. Were Dr Draco's theory of the State sufficiently authoritarian, he might hold that any crime is a form of *lèse-majesté*, and hence presumptively quite grave. But in the previous chapter, a different (and less authoritarian) theory of crime-seriousness has been suggested, according to which the gravity of offences would depend in important part on how much they typically affect a person's living standard. Low-ranking crimes only involve a small amount of harm, judged in terms of their impact on the typical victim's living standard, and hence are not worthy of much disapprobation.

Second, Draco's proposed penalties—imposing significant periods of incarceration even for those convicted of the least reprehensible acts— deprive the persons involved of important interests, namely, those of freedom of movement and association. Those interests are important because of their impact on the living standard of those punished.★

Third and finally, punished persons' vital interests are being trivialized, when such drastic deprivations are used to convey merely a mild degree of censure. By permitting periods of incarceration to be used as a symbol of lesser blame, the system conveys the message that the interests of those punished are of little importance. The message is that we are not much shocked by low-ranking offences, so we will 'only' take the offender's liberty away. This implies that his liberty—a matter of great importance to him—is of slight import as far as the State is concerned.

Whether cardinal proportionality is satisfied, however, is not so much a simple yes–no matter as a matter of degree. Consider a hypothetical penalty scale which still is harsh, but not quite so much so as Dr. Draco's. Imprisonment (of durations varying with the gravity of the crime) is prescribed for all intermediate and serious offences. Minor crimes do not receive imprisonment—but do nevertheless receive rather onerous non-custodial sanctions such as stiff fines. Does this scale violate cardinal-proportionality constraints? One would have to say that the scale raises questions comparable to the Draconian scale, but in a somewhat lesser degree. Thus the upper bound of cardinal proportionality would be illustrated not so much by a bright line as by a shaded area, that grows progressively darker as one is speaking of more drastic penalty schemes.

Are there comparable lower bounds of cardinal proportionality, restricting the extent to which a penalty scale may be deflated? I once assumed

★ See the discussion of gauging sanctions' severity in the preceding chapter.

there were.[1] The most serious offences on the scale—say, murder—call for an expression of severe disapproval. Such disapproval cannot adequately be conveyed if the scale is sufficiently deflated—say, so much so that murderers were only fined.

Nils Jareborg, I now think correctly, has challenged this conclusion.[2] In Chapter 2, I contended that the only reason for preferring the criminal sanction to a system of formal censure alone (or of censure accompanied by token deprivations) is the sanction's preventive function. If so, Jareborg argues, a drastic deflation of penalties would presumably be permissible, were that not to lead to unacceptable losses of prevention. Were such deflation carried far enough, one might eventually approach a system of censure with token impositions. Granted, more substantial penalty levels may be needed in our tougher, real world. But this means that it is preventive concerns that ultimately constrain a deflation of penalty levels, rather than considerations of cardinal desert.

2. ANCHORING THE SCALE WITHIN CARDINAL CONSTRAINTS

Since cardinal proportionality places only the outer constraint just described, much leeway remains for locating the scale's anchoring points. We thus need to identify what other factors are relevant.

Nigel Walker asserts that the very existence of this leeway vitiates a desert model. If the theory cannot establish actual levels of deserved punishments, he says, it is largely useless, and hence even its ordinal scaling requirements need not be taken seriously. Walker concludes that penalties should be ranked according to preventive and utilitarian principles.[3]

Walker's conclusion strikes me as patently fallacious. The implicit assumption is that proportionalism is valid only if it supplies answers to *all* questions of fixing penalties. But why should this be so? If the principle of proportionality tells us how penalties should be ordered, it is helpful for *that* purpose even if we would mainly have to look elsewhere in deciding where the scale should be anchored. But still, we need to press on with inquiring what those other anchoring principles are.

The Kleinig Formula

In his 1973 book *Punishment and Desert*,[4] John Kleinig comes up with a way of anchoring a penalty scale. The scale, he suggests, can be anchored at its upper end by the severest penalty that can humanely be inflicted (say, 25 years' imprisonment*) and at its lower end by the mildest penalty feasible

---

* Those limits would be based, not on notions of desert, but on the idea of humane and decent treatment of human beings. A specified number of years' imprisonment, such as 25 years, might constitute the upper limit. I am assuming that the death penalty is not humane or decent. For further discussion, see J. G. Murphy, 'Cruel and Unusual Punishment' (1979).

(say, a very small fine). Then the severest penalty would be applied to conduct having the highest seriousness-rating, and the mildest penalty to conduct having the lowest rating. With the scale thus anchored at each end, penalties for other crimes could be ranked and spaced in between to reflect the relative ranking and spacing of the criminal conduct involved.

Kleinig's scheme constitutes *a* way of fixing the penalty scale's anchoring points. It thus provides an answer to assertions that it is impossible to find a rational basis for deciding the magnitude of the penalty scale. His scheme does not, however, suggest the only or necessarily the best way of anchoring penalties. Different magnitudes could be generated by compressing Kleinig's scale either upward or downward. One could, for example, anchor the scale at the top with a penalty well below the maximum that could humanely be inflicted (say, 5 or 10 years instead of 25). Kleinig has not supplied reasons why such a compressed-downward (more lenient) scale should be rejected.

Penal Capacities as a Basis

In an earlier volume,[5] I suggested consulting the capacities of the penal system (particularly, prison capacities) as a way of helping to fix the scale's anchoring points—providing the results do not infringe cardinal-proportionality constraints. This can give practical guidance to a rule maker; and was, indeed, used by Minnesota's sentencing commission in the drafting of its guidelines.

Minnesota's commission sought to make penalties more proportionate, by reordering them so as to give greater emphasis to the seriousness of the crime (rather than the criminal record, which sentencing judges had chiefly looked to in the past). The commission decided that aggregate use of imprisonment under this reordered penalty scheme should not exceed existing prison capacity. Relying on prison capacity was thus employed to set the scale's anchoring points, and was quite helpful as a practical matter. Minnesota's politics prevented a significant scaling-down of penalties, and the guidelines' drafters were concerned primarily to prevent further overall increases. Relying on prison capacities served as a way of avoiding increases, and provided an incentive for a more realistic discussion of penalties.[6]

Prison capacity, however, cannot be a principled answer to the anchoring question. Braithwaite and Pettit are correct in saying that we should be troubled 'if, within a federal structure, punishments in State X are twice as severe as in State Y, simply because State X has a lot of spare cells'.[7] Relying on existing prison capacities seemed attractive in Minnesota because the state had (by American standards) used imprisonment rather sparingly. But in another state, with different traditions, reliance on existing penal capacities could have produced much higher penalty levels. And even in

Minnesota, one ultimately cannot avoid the question of whether the state acted appropriately in using imprisonment as much as it has. In short, the capacity of the penal system is a question of policy which should depend on how the penalty scale should be anchored, not vice versa.

A Decremental Strategy?

Why not adopt a desert theorist's version of the decremental strategy? Initially, the scale's magnitude would be chosen somewhat arbitrarily, say, on the basis of current penal capacities. Once thus chosen, rateable reductions in penalties would progressively be made. These would continue until further reductions would be deemed to sacrifice crime prevention unduly.*

How does this model differ from Braithwaite and Pettit's 'decremental strategy'?** One obvious difference is that penalties on the scale would be ranked according to crimes' seriousness. Decrementalism would be used to generate reductions of the scale as a whole, not selective reductions. In Braithwaite and Pettit's strategy, by contrast, ordinal proportionality would be ignored, and serious offences could end up receiving penalties below those for lesser ones.[8]

Might there not, however, be a second important difference, relating to the conception of prevention used? Penalties, according to the decremental strategy, are to be reduced until a floor is reached based on crime-prevention concerns. According to Braithwaite and Pettit, crime prevention—indeed, everything on their model—is to be based on what would promote citizens' safety and security (in their words, 'dominion') to the optimal aggregate extent. However, desert theorists—as was pointed out in Chapter 3—are not necessarily committed to any aggregate optimizing model at all. So might the 'floor' for the decremental strategy be based on some alternative notion of prevention?

The 'Optimizing' View of Prevention and its Difficulties

When penologists speak of crime prevention, they tend to lapse into classic penal utilitarianism. Optimum levels of prevention are to be achieved by weighing preventive benefits (aggregate crimes prevented) against aggregate costs (including the pain to those punished).[9] When it comes to the ordering of penalties, desert theorists have resisted this thinking—because they insist that penalties should be graded to reflect crimes' seriousness. But when the anchoring of the scale is considered, these grading requirements have been satisfied already. If prevention can then come into the picture, it is easy to

---

* Such an approach was sketched in my *Doing Justice*, (1976), ch. 16.
** See Ch. 3 for discussion of those authors' views.

revert to the classic model—to setting penalty levels so as to prevent crime efficiently.*

It would seem at first sight that such an optimizing approach to fixing anchoring points could be squared with a decremental strategy. In so far as anything is known about aggregate preventive effects, it appears that crime rates are not particularly sensitive to variations in punishment levels. This apparent insensitivity is suggested not only by available deterrence studies,[10] but in natural experiments. The most striking of the latter has been the fact that the extraordinary recent increase in penalty levels in the USA has had so little apparent impact on crime rates.[11] If aggregate criminality is, indeed, not influenced much by increases or decreases in penalty levels, then a quite substantial lowering of these levels could be accomplished without large increases in crime rates occurring.

We cannot be certain, however, that this would be the result. Research on prevention—for example, on general deterrence—has encountered great difficulty in assessing the marginal preventive effects achieved by varying punishments for particular offence categories.[12] In the present context, we are speaking not about specific kinds of offences, but about the possible preventive impact of reductions in the penalty scale as a whole—where the estimation problems will be more formidable still. Thus while I think it plausible to assert that penalties could be reduced considerably without notable increases in crime rates occurring, this cannot be asserted with confidence.

If we are not certain what the effects of altering penalty levels are, then we also cannot be sure that the optimizing conception of prevention would support a decremental strategy. As pointed out earlier,** it may not be clear that optimally effective crime prevention would call for a decrease in penalties until a significant increase in crime occurs, rather than the obverse strategy of increasing penalties until significant *reductions* in crime are achieved. Which approach is preferable, on the optimizing view, would depend on empirical estimates about how much additional prevention is achievable, if any, through altering penalty levels. Such uncertainties render the optimizing view of prevention a shaky basis for a decremental strategy.

There are, moreover, more fundamental doubts. Let me refer back to Chapter 2, where I speak of the justification of the existence of punishment. One of the issues discussed there is why punishment—with its element of hard treatment—is preferable to purely or primarily symbolic condemnation. My answer was that the blaming element in the criminal sanction conveys that the conduct is wrong; and the hard-treatment element supplies a person with a supplementary prudential reason for refraining from the

---

* In *Past or Future Crimes* (1985), I took this approach in my treatment of fixing anchoring points through use of 'categorial' incapacitation strategies. See ch. 13 of that volume.

** In Ashworth's and my critique of Braithwaite and Pettit, Ch. 3, above.

conduct. Were someone to concur with that explanation, he would be prepared to accept a system of sanctions that both expressed blame and included material disincentives designed to reduce his temptation to engage in the behaviour. Suppose that such a system could be embodied in a penalty scale with comparatively modest sanctions. The person, following this reasoning, accepts such a system, and concedes that the imposition of its sanctions on him would be proper were he to offend.

Now imagine that, instead of using these modest levels, the scale is anchored using the optimizing view of prevention. Suppose that doing so would sharply increase penalties pro rata on the scale. Those increases make the system more efficient, let us suppose, because its higher penalty levels would take the most active violators out of circulation for longer periods.[13] Were we to propose this revised scale to our friend, he might have reason to object that his vital interests are being sacrificed to the social good. It is one thing, he might say, to impose a moderate scale of sanctions that express the wrongfulness of the conduct, and also provide a practical disincentive not to engage in it. He might concede that, as a fallible human being, he might be tempted to offend and thus require a prudential reason for resisting the temptation. It is quite another thing to sacrifice his vital interests, should he offend nevertheless, as a tool for promoting the most socially efficient system of prevention.* Why, he would argue, should the penalties he and other ordinary persons face be increased so much simply because the resulting scale could deter or incapacitate a certain group of potential violators more efficiently?

A Revised View of Prevention?

Perhaps prevention can be viewed differently. Let me refer again to Chapter 2, where I sketched my two-pronged account of punishment, according to which the criminal sanction provides both a moral reason and a supplementary prudential reason for compliance. Such a view, I suggested, is respectful of the agency of the person. A responsible person—one capable of recognizing his obligation not to do injury—may still perceive himself as liable to temptation. If so, he could support a sanction system that also gives him pragmatic reason for not succumbing to the temptation.

Crucial to such a conception of prevention is that it be implemented through a modest range of sanctions. Such sanctions would provide a supplementary prudential incentive for compliance, which may help offset the temptation to offend. But since the threat levels are not so devastating, the moral reasons expressed through the sanction's censuring features still

* Our friend has reason to care about those higher penalty levels because disincentives, at almost any level of onerousness, cannot be expected to function infallibly. Even if a modest level of sanctions gives him and people like him a significant prudential reason to desist, he might still offend in a moment of weakness or wilfulness.

count. The actor would not suffer irretrievable damage to his life prospects were he to ignore the threat and suffer the penalty. In such a situation, the actor's response to the normative message of the sanction still has some practical relevance to his decision. The supplementary prudential disincentive is just that, supplementary: it does not loom so large as to co-opt or displace the normative message.

How might the penalty scale be anchored on this suggested view of prevention? Unavailable data—such as the data on deterrence called for by the optimizing view—would not be needed. It should be possible, using common sense, to discuss what overall levels of punitiveness would create a reasonable set of disincentives backing up a censure-based system. In this connection, it would be appropriate to compare those sanction levels with other normal setbacks and hard knocks people receive in their daily lives.

Concededly, such a conception is not precise enough to point to a particular appropriate level of sanctions. If a proposed magnitude is selected somewhat arbitrarily, however, it may be possible to argue whether it meets the suggested standard. Consider the scale I recommended a decade and half ago, in *Doing Justice*.[14] Incarceration, I proposed, was to be limited to serious offences (primarily, crimes of violence and the graver white-collar crimes); and durations of confinement for these offences should run up to three years—save for homicide, where five years would be the normal limit.* Such a scale is sharply below current English and U.S. levels.[15]** But if one applies the criterion I am suggesting here, namely, whether such a scale creates sanctions that would constitute a significant prudential reason for desisting, the answer is surely affirmative.

Let me also adduce an argument that is (in a larger sense) political. The conception of prevention I have sketched would seem suited to a responsible citizenry. Such persons would recognize their duty not to offend; would understand that they would sometimes nevertheless be tempted; and would opt for a range of moderate penalties sufficient to make a normally-fallible person seriously reconsider before offending. But it would not seek to *compel* compliance—since citizens would be expected to respond chiefly to sanctions' censuring message, plus a modest degree of prudence.

A democratic society, in my view, is one whose basic institutions should be designed with such a responsible citizenry in mind. This is most obviously true with institutions such as the franchise, but should be true of

* My proposal did permit sentences of over five years for homicides of the most heinous nature.

A problem with my proposal at the time was that I was not able to propose an adequate array of non-custodial penalties, including those of intermediate severity, for offences other than the serious crimes that would call for incarceration. I would now suggest non-custodial penalties of the kind discussed in Ch. 7, below.

** However some other European countries (Sweden, for example) rarely use imprisonment of over five years' actual duration.

punishment as well. Punishment may deal with the situations when citizens misconduct themselves, but even there the response should be one suited to rational, autonomous adults. The censure in punishment rests on the assumption that offenders are capable of appreciating the implicit message of disapprobation; the hard treatment should likewise rest on the assumption that offenders or potential offenders can weigh modest practical disincentives in deciding whether to offend. A penal system with these features treats the person as a self-governing adult even if he or she does consider offending.

The optimizing view of prevention is based on different—and, I think, troublesome—political assumptions. Potential offenders are seen as disturbers or potential disturbers of the social peace, whose depredations are to be prevented as effectively (or cost-beneficially) as possible. How repressive such a system would be would then depend chiefly on what works. This conception shows itself in rawest form in utilitarian sentencing schemes, such as selective incapacitation strategies.★ Applying cardinal and ordinal desert constraints may make the resulting system somewhat less harsh or unfair, but its basic political evil would persist: the creation of a major social institution that treats persons as something other than responsible subjects.

It may be objected that criminals, those with whom the criminal law must deal, scarcely fit the paradigm of the responsible citizen, for what leads them to offending may precisely be their weak sense of obligation to others. How can the criminal law be constructed with a picture of its subjects in mind which is so at variance with their actual nature? The short answer is that the crime-preventive mission of the criminal law is not necessarily that of preventing the most recalcitrant from offending. The criminal sanction cannot achieve anything near one hundred per cent efficiency, and fulfils its proper function if it helps induce most people to comply. My conception of prevention is one that would give ordinary persons good reasons for compliance.

Even if some prevention is thus achieved, might not some also be lost? Might not an optimizing preventive strategy promote public safety better? Possibly, yes—for the obvious reason that my alternative conception of prevention is designed with considerations in mind other than those purely of preventive efficiency. But the search for maximum efficiency has not yielded much to date, perhaps, because of the earlier-mentioned insensitivity of crime rates to variations (other than extreme ones) in punishment levels. And the other values implicit in my conception strike me as ones worth pursuit in a free society. There also remains the possibility, explored in the next chapter, of permitting exceptional departures from the normal penalty standards where extraordinary preventive losses are otherwise likely to occur.

★ See Ch. 6 below, and also my *Past or Future Crimes*, chs. 9–11.

Let me emphasize: the conception of prevention suggested here bears on the question of setting anchoring points—on how far the decremental strategy may properly be pursued before reaching a level of sanctions minimally needed for prevention. It does not deal with the grading of penalties, where the requirements of ordinal proportionality continue to apply.*

The Decremental Strategy Revisited

My revised conception of prevention should help clarify the decremental strategy. That strategy, we saw, was that of reducing penalties pro rata until a floor suggested by preventive considerations was reached. The difficulty with using the traditional 'optimizing' view of prevention, we saw also, is that it might possibly (even if not probably), result in a floor that is high—indeed, higher than existing penalty levels. In that event, the decremental strategy would become an incremental one, instead. My revised conception eliminates this risk. Once prevention is seen as a supplementary prudential disincentive, it would be achievable by rather modest sanction levels. The decremental strategy will, indeed, be decremental.

How, then, could the decremental strategy be implemented? One could begin with current severity levels, and move downward. Current levels would be those consistent with existing penal capacities and practices, but with penalties reordered to reflect the seriousness of offences. Since the censure expressed through punishment is a convention, these levels would reflect the present censure-expressing convention.

That, however, would only be the starting point. Conventions can be altered, and there is a powerful reason for changing existing censure-expressing conventions by moving the graded penalties downward pro rata. The reason is, of course, parsimony: keeping State-inflicted suffering to a minimum.

As penalties are thus reduced, the major constraint is that of practical politics. Reductions may well encounter resistance from important criminal-justice constituencies and from the public. How great that resistance is, and how far it will be possible to proceed with rateable reductions, will vary with the jurisdiction. Indeed, if crime is, or is perceived to be, on the rise, it will be difficult to proceed far with reductions, or even, perhaps, to restrain increases in penalty levels.**

Suppose, however, that considerable reductions were feasible. When should progressive diminutions in the penalty scale stop? Here, my revised conception of prevention points to a solution, albeit an imprecise one. Penalty levels should be reduced until the system provides the rather

---

* See Ch. 2 for the reasons why.
** See Ch. 10 for further discussion of such political problems.

modest disincentives of which I have just spoken. An example would be the up-to-five-year scale referred to above.

'Utopianism' and Practical Relevance

Am I not being utopian here? Not many jurisdictions are likely in the near future to undertake the rather drastic penalty reductions that I suggest. If they are not, what use is my proposed version of decrementalism?

One use it has is to indicate the direction sentencing policy should take. Maybe sentences cannot, in most places, be reduced down to the rather low levels I propose. But, for the reasons explained, the direction of sentence levels should be downward. My proposed decrementalism thus constitutes an ideal we should be seeking.

Another relevance of my proposed decrementalism is to answer a commonly-cited objection to desert theory. A proportionalist conception of sentencing, it is asserted, can tell us nothing about where the anchoring points of the scale can be fixed, within cardinal-desert constraints. That conception thus is said to be consistent with rather harsh sentencing policies. I think I have answered this objection. A desert advocate does not have to be silent on the question of anchoring points. A theory can be articulated—as I have attempted in this chapter—that supports and accounts for a decrementalist strategy. The theory's conclusions are not *compelled* by the logic of desert, for we are speaking of anchoring-point decisions that are to be made after cardinal- and ordinal-proportionality constraints have already been satisfied. However, the rationale for proportionality set forth in Chapter 2, and the anchoring-points theory suggested in the present chapter, have a common basis in the conception of censure and prevention in the criminal law that I have been sketching.

6

Hybrid Models

A DESERT-BASED sentencing scheme is somewhat confining, in its requirement that offence seriousness—and not a variety of other factors—should chiefly decide comparative punishments. Its confining character facilitates scaling penalties in a coherent fashion, but it also limits possibilities of seeking other goals and objectives. Why, however, might not other values override those of proportionality, at least to some extent in some situations? We need to examine 'hybrid' models that allow a degree of departure from ordinal desert requirements. Even a desert model permits reliance on other factors (including those of crime prevention) for some purposes—e.g. for deciding among sanctions of comparable onerousness.★ When proportionality constraints are thus satisfied, merely seeking subsidiary goals does not render a scheme a hybrid one. A hybrid *departs* from the assumed primary guiding principle—in this case, proportionality—to achieve ulterior objectives.[1]

I shall be speaking here of hybrids of a certain sort: those that permit only a limited degree of departure from the principle of proportionality. If proportionality is an important fairness constraint, then it ordinarily should restrict the pursuit of ulterior ends, such as those of crime prevention. Hybrids that depart from proportionality in restricted situations or to a restricted extent still would qualify as broadly equitable schemes. I shall not be considering models which routinely disregard proportionality, or treat it as only a marginal constraint.

The reader will detect a certain tentativeness here. The subject of desert-based hybrids has received comparatively little attention. It is not easy to construct a theory that allows one to decide when proportionality may be trumped by other goals. What makes matters harder is that the hybrids of which I speak would depart from proportionality only to *some* extent: on one hand, one cannot readily dismiss such schemes as being basically unfair; on the other, it is not easy to discern how much ulterior benefit may be obtained from restricted departures. My discussion will thus be exploratory: to see how one might conceptualize hybrid models. I remain sceptical of these models for reasons to be explained, but others may see their value differently. The point of examining these hybrids is to make it clear that a

---

★ See the discussion of interchanges among penalties in Ch. 7, below.

commitment to proportionality does not rule out pursuit of other objectives: that we may agree that sanctions should be proportionate in the main, and still debate a certain degree of deviation.

1. PROPORTIONATE SANCTIONS WITH EXCEPTIONAL DEPARTURES: PAUL ROBINSON'S HYBRID

In examining hybrid schemes, it might be best to begin with that suggested by Paul Robinson in a thoughtful 1987 essay.[2] Under his scheme, penalties ought ordinarily be scaled according to crimes' seriousness, as required by the principle of proportionality. Upward deviations from ordinal desert requirements would be permitted, however, in exceptional circumstances—if needed to prevent an 'intolerable level of crime'. However, Robinson would impose a further limitation on such departures: that even when the prevention of major criminal harm is at issue, *gross* deviations from proportionality would not be allowed.

Robinson couches his formula in general terms, terms such as preventing an 'intolerable' increase in crime. What is or is not tolerable is a question of judgement, and Robinson is not so much offering a criterion as a way of thinking about departures. However, the following examples[3] illustrate how Robinson's model might be applied.

1. Sweden's 1988 sentencing statute[4] calls for sentences to be based chiefly on the gravity of the criminal conduct. However, an exception was made that had the effect, *inter alia*, of permitting continuation of Sweden's long-standing policy of imposing short jail terms on drivers (including first offenders) found driving with more than a stated blood alcohol level.[5] This policy was designed as a deterrent, and constitutes a departure from the statute's general sentencing structure. (Ordinarily, persons convicted of offences in the upper-middle range of seriousness—which is how this offence would be rated* —would receive a non-custodial sanction, unless they had accumulated a significant criminal record.) The departure might arguably (but with reservations discussed later) be defended on Robinson's theory, because drinking and driving, in the aggregate, wreaks so much serious harm. The amount of deviation from proportionality is also not very great, for offenders would receive a short period of confinement in lieu of the somewhat less onerous intermediate penalties they would otherwise receive.

2. US advocates of 'selective incapacitation'[6] have proposed giving convicted robbers who are classified as high risks lengthy extensions of their prison terms: as much as eight years' imprisonment for allegedly

---

* If the drinking driver actually conducts his vehicle in a hazardous fashion, he would be guilty of the more serious charge of reckless driving. The drinking-and-driving offence is regarded as being in the intermediate gravity range, because no other person may actually have been endangered by the defendant's conduct in the circumstances.

high-risk robbers, as contrasted with as little as one year's confinement for the lower-risk ones. Large preventive benefits have been claimed for such a strategy, although those claims are now disputed.[7] Smaller differentials in sentence between high-risk and lower-risk robbers would cause the projected preventive effects to disappear. Even if such a policy had the preventive benefits its advocates claim, it would be questionable under Robinson's model, because it would routinely involve such drastic penalty disparities on grounds wholly ulterior to the gravity of the conduct. That would seem to constitute the kind of gross infringement of proportionality that is ruled out under the Robinson scheme.

How can one justify Robinson's model? A simple argument would be to invoke the analogy of quarantining persons with deadly and easily communicable diseases.[8] Quarantined persons surely do not deserve to lose their liberty, for it is not their fault that they are disease carriers. They are deprived of their freedom solely in order to protect the health of others. The reason for tolerating quarantine is that community survival is deemed paramount to concerns about justice.

The quarantine parallel, however, is not persuasive when given closer scrutiny. The criminal harm at issue is not comparable in magnitude to the injuriousness of epidemic diseases: drinking and driving, for example, represents no such threat to community survival as cholera or plague epidemics. Gauging the preventive effects of additional penalties also is scarcely as reliable as, say, diagnosing typhoid carriers.

Punishment, unlike quarantine, involves blaming. The person who receives extra punishment suffers censure to an extent not warranted by the blameworthiness of his or her conduct. We would not tolerate treating quarantined individuals as bad persons who *deserved* confinement. The quarantine analogy is also too ambitious. A quarantined person may be detained indefinitely, without regard to fault, as long as he remains a disease carrier. Robinson's model, however, would only ease desert constraints in exceptional situations, not eliminate them—because of his limit that the extra punishment must itself not be 'intolerably unjust'.

If not quarantine, what other argument can be made? Here, Ronald Dworkin's model of rights is helpful.[9] Dworkin[10] maintains that rights constitute claims against the general welfare: if someone has a right, it should be observed even if doing so would not be in the societal interest. Rights, however, are only *prima facie* claims: they may sometimes be overridden when the countervailing concerns are of sufficient urgency. One overriding ground, Dworkin suggests, is when the loss of utility involved in maintaining the right is of extraordinary dimensions: when, in his words, 'the cost to society would not simply be incremental, but would be of a degree beyond the [social] cost paid to grant the original right, a degree great enough to justify whatever assault on dignity or equality might be involved'.[11]

Whether all rights can be described in Dworkin's fashion, as claims against the general welfare, has been disputed among philosophers.[12] That issue, however, need not concern us here. Whatever the general logical structure of rights may be, the proportionality principle does function as a fairness-constraint which limits pursuit of social goals—particularly those of crime prevention. That constraint might apply in ordinary circumstances, and yet yield when the resulting loss of utility—of crime prevention—would be of unusually large dimensions. Dworkin's formula, therefore, seems a plausible enough way of conceptualizing this hybrid model.

There are, nevertheless, a number of key questions that Dworkin's model does not resolve and that need to be addressed. One concerns the strength of the *prima facie* equity demand. Some fairness constraints should be less readily overridden than others. Consider the requirement of proof beyond reasonable doubt: it would scarcely seem proper to permit *that* requirement to be trumped, even in order to avoid major losses of crime prevention.[13] What Robinson must implicitly be claiming, therefore, is that proportionality is an important fairness-constraint (important enough to require major losses of utility before it may be overridden), but not *so* crucial as the proof-beyond-reasonable-doubt requirement (for which no such override would seem justifiable). Another question concerns the weight needed for the overriding-utility claim. What is involved is not community survival, as was the case with quarantine. However, Robinson seems to suggest that the criminality sought to be prevented through the override must involve both substantial frequency and a high degree of injuriousness, e.g. a substantial increase in the incidence of conduct threatening life and limb.[14]

One is left largely with intuitions on these questions, because they involve apples-and-oranges comparisons. One has, on one hand, a non-consequentialist, retrospectively-oriented fairness demand; and on the other, an override claim based on consequentialist notions of future harm. No single intellectual currency exists into which these competing considerations can be translated and compared. Perhaps the case for Robinson's solution is simply that it makes proportionality not an absolute constraint, but an important one. The countervailing utilities must be weighty for an override to be warranted. Moreover, the extent of permitted deviation from proportionality is also restricted. Gross departures would be impermissible—because such manifestly disproportionate responses would misrepresent wholly the degree of the person's blameworthiness.[15]*

The Robinson hybrid has its undeniable appeal. While abiding by desert

---

* This latter constraint explains why the departure should, on Robinson's model, consist of an extension of the sentence—and not of a separately-imposed period of civil commitment, to be served after the normal sentence. The latter (because not denominated as punishment at all) could be deemed to be exempt altogether from from the proportionality principle. The extra confinement thus might be of indefinite duration.

constraints ordinarily, it permits departures where the case for them seems the strongest. What, then, are the potential problems?

One problem concerns the sufficiency of evidence to override. Robinson would restrict departures from proportionality, as noted, to cases where the conduct is not only very harmful to its victims, but a significant incidence of the conduct is involved. In the drinking-and-driving example, many potential victims, not merely the occasional one, may be affected by this type of conduct. This, however, sharply restricts the scope of the exception, since there so seldom exist sound empirical grounds for believing that a departure from desert requirements would significantly reduce the aggregate incidence of the conduct. The desired crime-reduction effect would derive from general deterrence or incapacitation: either actors are intimidated better through the extra penalty or that penalty provides added restraint. It is, however, notoriously difficult to confirm such effects. Consider the drinking-and-driving example, again. While there is reason to believe that Sweden's policy of penalizing drinking and driving has had some preventive impact, it is far from clear that significant extra prevention is achieved by routine resort to imprisonment even for first offenders.[16] It is thus worthy of note that in 1990, Sweden in effect narrowed the presumptive-imprisonment exception, so that a considerably smaller number of drinking drivers now go to prison.[17]

Were overall penalty-levels reduced as I propose in the previous chapter, the problem of exceptional losses of prevention for certain types of crime might come into sharper relief. But even then, it will not be easy to show that such losses are, indeed, occurring.

If the requirement of traceable *aggregate* effects presents these difficulties, might it be dropped? A. E. Bottoms and Roger Brownsword have suggested doing so: individuals who constitute a 'vivid danger' of seriously injuring others should be given a period of extra confinement, even if such a policy has no measurable impact on aggregate violence levels.[18]* However, these authors emphasize that such an exception should be invoked only when there is a high and immediate likelihood of serious injury occurring.

Even this latter standard of individual dangerousness, however, is not

---

* Dworkin suggested two possible grounds for overriding a right. One, referred to in the text, is where a loss of utility of extraordinary dimensions would otherwise occur. Another is where a competing right is involved, e.g. when rights of free speech collide with those of privacy. In my view, it is the first of these two grounds which is relevant here: the issue is that of overriding proportionality because of the extraordinary hazard posed by certain high-risk offenders. Strangely, however, Bottoms and Brownsword assert that it is the second of Dworkin's two grounds that is involved: the offender's claim to not-more-than-proportionate punishment is defeated by the potential victim's claim not to be injured. This I find puzzling. Granted, the victim has rights of physical safety which the offender infringes. But has *the State* infringed the victim's rights if it merely fails, as law enforcement so often does, to prevent the injury? Suppose, particularly, that the State responds to the offender's current crime by imposing a proportionate sanction. It is then far from clear why potential victims have a *right* against the State to a more stringent sanction that might possibly forestall future victimization.

easily met. To justify extension of the sentence on grounds of Bottom's and Brownsword's 'vivid danger' criterion, the person would have to be likely to injure others even *after* expiration of his normally-deserved term for the crime of conviction. Making such a forecast presupposes a capacity to gauge the expected length of a criminal career—a capacity that does not exist today.[19] The uncertainty of career durations also may mean that lengthy extra confinements may be needed to provide any assurance of prevention. The longer the confinement, however, the more it collides with Robinson's restraining principle that grossly disproportionate sanctions are to be avoided. Thus someone who concurred in principle with the idea of extending sentences in cases of 'vivid danger' might still doubt our present capacity to assess that danger with the requisite assurance.

Another potential hazard is that of erosion of the standard. Critical to Robinson's (as well as Bottoms's and Brownsword's) model is a narrow departure criterion: that the sentencer may impose a more-severe-than-proportionate sanction only to prevent grievous criminal harm. Loosening that standard, so as to admit lesser harms, compromises the basic idea of the hybrid: that desert constraints (as important requirements of justice) may be disregarded only in exigent circumstances. Yet how confident can one be, given the political dynamics of crime legislation in most jurisdictions, that narrow departure standards could be maintained? May not a narrow exception be expanded too easily in the name of protection of the public? Someone might support this hybrid model in theory *if* its tightly-drawn departure criteria could be sustained, and still be worried about implementing the model because of that 'if'.[20]★

Robinson's model, as well as Bottoms's and Brownsword's variant, deals with exceptional increases in sentence. What about exceptional *decreases*? The logic could be similar: deviating downward, when necessary to secure sufficiently urgent ulterior ends. The problem is identifying those ends. Crime-prevention is not a plausible candidate, as it is difficult to imagine extraordinary losses of prevention stemming from a failure to lower the sentence for a particular type of crime below its proportionate level. Parsimony is not a plausible candidate either: exceptional downward departures (and exceptions is what the model contemplates) will not have much of an impact on overall sentence levels. There might, however, be

---

★ England's Criminal Justice Act of 1991 provides an interesting test case. While the statute ordinarily makes proportionality the criterion for determining sentence, an exception permits invoking confinement for dangerous offenders. The exception appears intended to be narrow: confinement-for-dangerousness can be imposed only for offenders convicted of certain offences, and then only when needed to protect the public from 'serious harm'. Serious harm is defined as 'death or serious personal injury, whether physical or psychological'. However, the likelihood of that harm is not addressed—Bottoms's and Brownsword's suggested criterion of immediacy and high likelihood thus is not included. Serious psychological injury may also prove to be too flexible a term. It will be interesting to see how broadly or narrowly the courts interpret these provisions, and also to see whether Parliament will expand its scope in the future.

other possible grounds. A difficult issue is whether administrative considerations—notably, co-operation with the prosecution under specified circumstances—should be a basis for downward departures.* Departures might also be invoked for certain other types of cases which appear to deserve special sympathy.** In any event, these selective penalty-reductions would call for a cogent explanation of what is to be accomplished, in what kinds of exceptional situations, and why that aim is urgent enough to override desert constraints.

2. 'RANGE' MODELS

Robinson's hybrid allows departures from ordinal desert only in special situations. An alternative would be to allow relaxation of ordinal desert constraints more routinely. Desert considerations would be treated as setting an applicable range of punishments, but within that range the penalty could be fixed on consequentialist grounds. There are two different versions of such a model, having different rationales and different practical implications. Let us consider each.

'Limiting Retributivism'

This view is identified with the writings of Norval Morris.[21] Desert, according to his suggested model, is to be treated as defining only outer limits: not more than so much punishment for a given type of offence, nor (perhaps) less than so little. Within the resulting broad ranges, the sentence is to be fixed by reference to other reasons. Some German penologists have urged a comparable view, termed the 'Spielraumtheorie'.[22]

In some passages, Morris suggests that his model is required by the logic of desert. Proportionality, supposedly, is indeterminate: it suggests only how much is *un*deserved in the sense of being excessive or manifestly too lenient. Within these bounds, reliance on non-desert grounds is appropriate because the claims of desert have been exhausted.[23]

The difficulty of this view is that it overlooks the requirements of ordinal proportionality—particularly the requirement of parity.[24] When two defendants commit comparably culpable robberies, giving one a larger sentence than the other for the sake of (say) crime prevention visits more blame on one for conduct that is *ex hypothesi* equally reprehensible. To assert that desert, by its very logical structure, imposes mere limits disregards this demand for parity.

In other passages, Morris seems to concede that desert supplies not only

* For discussion of some of the problems of creating an explicit discount for such co-operation, see A. von Hirsch, K. Knapp, and M. Tonry, *The Sentencing Commission and Its Guidelines* (1987), ch. 9; see also, A. Ashworth, *Sentencing and Criminal Justice* (1992), 129–33.
** This possibility is touched upon in the Epilogue, below.

outer bounds, but also ordering principles including a parity requirement. This latter requirement, however, is said to be weak and easily overridden: it is no more than a 'guiding principle'.[25] Parity concerns may thus be trumped by competing values—some preventive (deterrence, incapacitation), and some humanitarian (reduction of penal suffering). This version, however, ignores the reasons (set forth in Chapter 2) why there is something important, not just marginal, about punishing similarly those who have committed comparably-serious offences. Abandoning or watering-down comparative desert requirements also leaves one without much guidance on how the only apparently-remaining desert constraints— Morris's supposed desert limits—are to be ascertained. Not surprisingly, neither Morris nor the German advocates of the 'Spielraumtheorie' have been able to suggest, even in principle, how those limits might be located.[26] It is thus unclear whether 'limiting retributivism' would make desert or something else (notably, crime prevention) the principal determinant of the sentence.

'Range' Models Recast as a Hybrid

There is, however, another way of conceptualizing a 'range' model: one that would make it explicitly a hybrid theory of the kind of which we are speaking. This version concedes that ordinal proportionality *does* require comparably-severe punishments for comparably reprehensible conduct— and thus that unequal punishment involves a sacrifice of equity. The extent of that sacrifice, however, depends on how great the inequality is. Why not, then, allow preventive (or other consequentialist) considerations to override desert, but only within specified, fairly modest limits? Variations in punishment for a given offence would be countenanced, provided the specified limits are not exceeded. The idea is to permit the pursuit of those other objectives without 'too much' unfairness.

This model differs from Morris's 'limiting retributivism' in that it requires closer scrutiny of inequalities in punishment. Since parity is regarded as an important constraint, not just one of marginal significance, it matters how much deviation from parity is involved—and for how strong ulterior reasons. Only modest deviations, to achieve other pressing objectives, would be permissible.

Under such a model, two major questions arise. The first concerns specifying the limits: how much variation from parity is to be permitted? The second is the identification of the ulterior ends: for what goal (crime prevention or what else) should such variations be warranted? Let me consider each of these issues, in turn.

Specifying the Limits

One objection to 'limiting retributivism', just noted, has been the difficulty of delineating the applicable desert limits. On this alternative model, the fixing of limits becomes conceptually easier. A specified degree of deviation from ordinal desert constraints is simply set as the applicable limit. Since the governing idea is that there should be only modest derogations of ordinal desert, those limits should be reasonably constrained. Perhaps a 10 or 15 per cent deviation might be permissible, but not a 50 or 60 per cent one. The gravity of the criminal conduct would thus continue substantially to shape (albeit not fully to determine) the gradation of punishment-severity.

Identifying the Ulterior Ends

For what purpose should such moderate deviations from ordinal desert be allowed? In the literature to date, the purpose usually mentioned is that of crime control—particularly incapacitation.[27] Applying this to the hybrid would mean reliance on offender risk would be appropriate, provided the applicable limits on deviation from ordinal desert were not exceeded. The end, then, is enhanced crime-prevention.

This strategy encounters, however, a fairness-effectiveness dilemma. A substantial incapacitative effect tends to be achievable only when the sentence differential between lower- and high-risk individuals is large.[28] Large differentials, however, mean infringing ordinal desert constraints to a great degree—and not the limited degree which the hybrid contemplates. Keeping the differentials modest, on the other hand, means restricted preventive benefits. That, in turn, raises a further difficulty: if ordinal proportionality is a demand of fairness, even limited deviations become justifiable only by a showing of strong countervailing reasons. How can this requirement be satisfied by a mere slight increase in preventive effectiveness?

Another possible ulterior end is that of parsimony. Advocates of 'limiting retributivism' claim that relaxing desert constraints permits milder sanctions.[29] Suppose that penalties are scaled on a desert model, and that the prescribed penalty for a given (fairly serious) offence is x months' imprisonment. Suppose one were to allow (say) a 10 per cent penalty reduction for non-dangerous offenders. Thus, some high-risk offenders will receive x months, and the remainder only 90 per cent of that period. Has not a penalty reduction been achieved?

The alternative, however, is to reset downward the anchoring points of the scale—so that all penalties would be rateably reduced by the 10 per cent.* Then, proportionality is not sacrificed. Moreover, parsimony is better achieved—because all (not just some) of those convicted of the crime get the benefit of lesser penalties. So why cannot one have parsimony while adhering to, rather than departing from, desert principles?

---

* On the anchoring-points theory sketched in the preceding chapter, such across-the-board reductions would be permissible until sanctions have been reduced to quite modest levels.

One possible response is that across-the-board reductions would increase the risk to the public, because even the high-risk individuals would have shorter periods of confinement. This, however, would turn the basis for the departure into the problematical one of crime-prevention, just discussed.

Alternatively, political considerations might be invoked: rule-makers in most jurisdictions, supposedly, will be more willing to accept penalty reductions for non-dangerous offenders than across-the-board reductions. If one is talking politics, however, is it realistic to expect the deviations to be in a downward direction only? The demand is likely to be not merely for less punishment for the low risks, but also *added* punishments for the supposed high risks. Then, the strategy becomes the rather less attractive one of less punishment for some offenders but more for others.

Selectiveness also diminishes the parsimony that is achieved. A 10 per cent reduction in overall punishment levels would represent a significant achievement, in terms of diminished aggregate suffering. (Indeed, such a reduction would be a significant first step in the decremental strategy discussed in the preceding chapter.) A comparable reduction only for a selected group accomplishes less: fewer persons are involved, and those receive only a modest diminution of their penalties. Permitting larger selected discounts would break the bounds of this hybrid model.[30]

What other reasons might there be for deviating from desert constraints? One might be to facilitate the scaling of non-custodial penalties. Under a desert model, substitution among sanctions would be permitted only when the penalties are about equally onerous. Relaxing proportionality constraints a bit might allow substitutions to be made more easily, and might also facilitate devising back-up sanctions for offenders who violate the terms of their sentence. This might be accomplished, moreover, while permitting only quite limited deviations from parity. The feasibility of such an approach will be examined more fully in the next chapter.

Where does this leave us? The attraction of the hybrids discussed in this chapter lies in the fact that they would seem to promote other goals at only small sacrifices of equity. Why not deviate from proportionality to only a limited extent, or only in restricted situations, if doing so would create a more effective or milder sanctioning system? The achievement of these ulterior goals, however, tends to be elusive. When scenarios in which these hybrids might be employed are examined carefully, it is seldom easy to show that much extra prevention or parsimony can really be accomplished. The focus on these other goals, moreover, can too easily lead to breaking the rather imprecise and fragile limits of these hybrid models. If restricted deviations from parity do not suffice to achieve the ulterior aim, why not larger deviations? Large deviations, however, mean a system that is no longer even approximately equitable. Appealing as these hybrids may seem, one needs to read the fine print.

7

Intermediate Sanctions

IN most English-speaking countries, sanctions have run to extremes. Either the offender has been incarcerated, or else placed on probation in a large caseload, with perfunctory supervision. Intermediate sanctions—those in the middle range of severity*—have largely been lacking. Since the mid-1980s, interest in intermediate sanctions has grown. A variety of such sanctions have been tried: day (or 'unit') fines based on the offender's income; community service; 'intensive' supervision; compulsory day or evening attendance at community facilities; and home detention.[1] However, these penalties were fashioned *ad hoc*, and applied to whatever heterogeneous groups of offenders seemed convenient. When new options were made available, it was believed, judges would make use of them instead of imprisoning. The strategy of simply creating additional options has proven disappointing, however. In the absence of principles governing their use, the novel sanctions have not necessarily been utilized as replacements for imprisonment. Instead, judges often sentence people to prison as before, and employ the new sanctions as substitutes for traditional non-custodial measures.[2] Sentencing guidelines might target the measures better, but those guidelines that exist in the USA generally deal only with the use of imprisonment and not with non-custodial sanctions.[3]**

There are now signs of change. Standards for intermediate sanctions are beginning to be developed. England's new Criminal Justice Act, for example, provides that proportionality requirements are to be applied not only to decisions concerning imprisonment, but also to the choice and imposition of non-custodial sanctions.[4]

Theory is just beginning to develop on how intermediate sanctions might be scaled. A 1989 article by Martin Wasik, Judith Greene, and myself

---

* 'Intermediate sanctions' refers to sanctions in the middle range of severity—including non-custodial sanctions more substantial than small fines or routine probation. Short stints of confinement may also be included, as discussed below. The term 'non-custodial penalty' has an overlapping but somewhat different meaning: it refers to any penalty not involving full custody in an institution of confinement. Non-custodial penalties thus include most kinds of intermediate sanctions (except short stints of confinement), but include lesser punishments (such as small fines) as well.

** The Minnesota guidelines, for example, address only the use of imprisonment in state correctional institutions. The guidelines do not specify which kinds of sanctions should be imposed, when imprisonment is not recommended.

attempts to grade such penalties according to a desert rationale.[5] A 1990 volume by Norval Morris and Michael Tonry offers scaling principles based on a mixed set of objectives.[6] Both our scheme and that of Morris and Tonry recognize that non-custodial penalties must be considered sanctions in their own right, not mere 'alternatives' to custody. Both recognize that such sanctions are *punishments*, involving both the deprivation and censure that characterize a punitive response. Both favour the development of explicit standards for the use of intermediate penalties. In this chapter, I wish to examine some of the issues that these schemes raise—issues concerning comparability and substitution among penalties, back-up sanctions for breach, and related matters.

1. DESERT-BASED SCALING

Desert can serve as the guiding principle in scaling non-custodial penalties. The von Hirsch–Wasik–Greene proposal[7] follows this approach, with the following elements resulting:

1. Non-custodial sanctions would be graded chiefly according to the seriousness of the crime of conviction. Intermediate sanctions, being by definition in middle range of severity, would be employed for crimes of medium and upper-medium seriousness. Lesser crimes would receive milder responses.

2. Substitution would be permitted among sanctions of comparable onerousness, but with policy-based limitations (discussed below) on how extensive that substitution may be.

3. There would be significant restrictions on the severity of the back-up sanctions that may be used against offenders who violate the conditions of a non-custodial sanction. Imprisonment could be employed only exceptionally.

Such a scheme has a number of advantages. First, it can restrict the use of imprisonment. Incarceration, as a severe punishment, would be prescribed for crimes of a more serious nature.* Offences of intermediate or lesser seriousness would generally draw non-custodial sanctions. Because such offences (for example, routine thefts, if repeatedly committed) are so often punished at present by imprisonment, the resulting reduction in use of the prison sanction could be considerable.[8]**

---

* The one possible exception, as we shall see below, relates to very short stints of confinement used as a substitute for certain intermediate sanctions.

** The extent of the reduction in use of imprisonment would depend on how the penalty scale is anchored, as discussed in Ch. 5. It would depend, also, on the extent to which imprisonment is currently used to punish offenders convicted of repetitive property offences. A desert scheme, as the discussion above notes, will tend to preclude imprisonment in such cases.

Second, the scheme would limit the use of intermediate sanctions (for example, those depriving the offender of significant portions of his income or leisure time) to crimes that are of middle-level gravity—and not permit their use for low-level offences, merely because their perpetrators seem more co-operative with programme requirements. It is the defendant who commits an ordinary burglary or auto theft, not the petty thief, who should be the prime candidate for such sanctions.

The scale of punishments could be reasonably simple, because of the limits on interchangeability. For each level of sanction severity, one type of sanction might normally be recommended. Substitutions of sanctions of equivalent onerousness would be permitted when there were special reasons—say, the defendant's inability to perform the tasks of the normally-recommended punishment. Free substitution would be barred, as would be the piling up of multiple sanctions on particular defendants.

Finally, the revocation sanction would be brought under control. It is a disturbing fact today that, in large penal systems such as those of Texas or California, a substantial percentage of those committed to prison enter via parole or probation revocation.[9] Intermediate sanctions could aggravate this tendency, if imprisonment is routinely resorted to as the breach sanction. Under a desert rationale, imprisonment may be invoked only for the quite serious breaches: those involving conduct that is comparable (or nearly so) in gravity to crimes warranting incarceration in the first place.

Each of these various features merits closer examination. Let me begin with the question of substitution among penalties.

2. INTERCHANGES: EQUIVALENT PENAL BITE

In fashioning guidelines for use of imprisonment, the question of substitutability of sanctions seldom arose, because imprisonment (at least for lengthier periods) seemed so much tougher than other penalties. With non-custodial sanctions, the problem presents itself in earnest: several comparable penalties may exist, and it is necessary to decide which should be invoked and under what circumstances.

The criterion for substitutions should, on a desert model, be that of comparable severity: approximate equivalence in penal bite. The principle of proportionality addresses the *severity* of penalties, not their particular mode. When sanctions are ordered according to the gravity of crimes, replacing a particular penalty with one of having roughly the same degree of onerousness disturbs neither the parity nor the rank-ordering of penal responses. Thus a day-fine sentence of so many days' earnings might be replaced by an equivalent stint of community service.[10]

Employing this criterion, of approximate equivalence in severity, requires judging the relative onerousness of the various penalties involved: a fine of how many days earnings, for example, is 'worth' a day of community service, or a day in home detention? Severity is, I suggested in Chapter 4, a matter of how much a sanction intrudes upon the interests a person typically needs to live a good life. Gauging comparative severity thus would involve comparing how much the various sanctions typically would intrude on a person's living-standard.*

What of crime-prevention concerns? A desert rationale does not permit crime prevention to be relied upon to decide among penalties of substantially differing severity, as that would infringe ordinal proportionality (particularly, the requirement of parity). That is the objection to schemes such as selective incapacitation, wherein prison terms of markedly different lengths are to be imposed on offenders, depending on their degree of risk.[11] When two kinds of penalties have approximately the same penal bite, however, parity is satisfied—in which event one penalty may be chosen over the other on preventive grounds.

A. E. Bottoms provides an illustration.[12] Suppose two offenders, A and B, are convicted of comparable offences of upper-intermediate seriousness. A is a first offender, and has a fairly 'stable' background. The other, B, has a criminal record. Since a desert model (arguably) allows the lack of prior convictions to be treated as extenuating to a limited extent,** A deserves a somewhat (but not much) reduced penalty compared to B. The record, however, is also an indicator of risk, and how should this be taken into account? Risk would not, on a desert rationale, justify a wider severity-differential than the quite modest one just indicated. However, risk may be used to decide the mode of the penalty: whereas A might receive a financial penalty, B's sanction might involve supervision of some kind, designed to reduce (at least partially) his likelihood of reoffending.***

---

* Such an analysis, examining the living-standard impacts of various intermediate sanctions, has yet to be undertaken. Meanwhile, however, severity comparisons could be made with the aid of common sense. Penalties could be classified according to the degree to which they affect a punished actor's freedom of movement, earning capacity, and so forth. This would permit rough-and-ready judgements to be made about the seeming 'weight' of those interests to a tolerable existence. Consider the comparative onerousness of day fines and community service. Both sanctions are economic (the latter because it exacts unpaid service), but community service also restricts movement by requiring the work to be done at a particular place. To take that difference into account, a day fine could be considered as equivalent to a given fraction of a working day in community service—that fraction being based on the rater's assumptions concerning the relative importance of those interests. The resulting rankings will reflect the rater's judgement, but will at least be supportable by explicitly stated reasons. See, M. Schiff, *Gauging the Intensity of Criminal Penalties* (1992).

** Some desert theories, such as Richard Singer and George Fletcher, have maintained that the presence or absence of a prior conviction is irrelevant to the offender's deserts. However, others—including Martin Wasik and myself—support a limited penalty-discount for the first offender, as a way of recognizing human fallibility in the criteria for deserved punishment. For fuller discussion, see A. von Hirsch, *Past or Future Crimes* (1985), ch. 7; M. Wasik, 'Guidance, Guidelines, and Criminal Record' (1987); A. von Hirsch, 'Criminal Record Rides Again' (1991), and A. Ashworth, *Sentencing and Criminal Justice* (1992), ch. 6.

*** This assumes that B's supervision sentence is slightly severer than A's financial sanction, but no more than that. It also assumes that the sanction chosen for B, the higher risk

Sentencing guidelines in the USA have tended to have a firm demarcation between custodial and non-custodial sanctions.[13] The sentencing grid has an 'in/out' line across it, with significant periods of custody prescribed for the (more serious) conduct above the line, and non-custodial sanctions prescribed or permitted for the (lesser) conduct below it. Any such 'hard' demarcation has its drawbacks: it can, for example, mean unnecessarily sharp transitions in severity as one moves from across the line, from non-custodial measures to custody.[14]

A desert rationale requires no sharp demarcation, because brief periods of incarceration may be equivalent in severity to the more substantial non-custodial impositions. Where this equivalence exists, substitutions could be permitted. The penalty scale might in its high ranges prescribe various durations of incarceration in excess of say, four months, without permitted substitution. The next most severe range of penalties might consist of fairly onerous community sanctions, such as home detention. Here, however, short periods of full custody—measured in days or weeks—would be the permitted substitute. The substitute might be invoked, for example, for offenders whose previous records of absconding suggests they would not remain at home as required by a home-detention order. The duration of that full custody would be calibrated so as to be comparable in its severity to the normally applicable non-custodial sentence.[15] Any such authority to interchange short prison terms with other sanctions, I should also emphasize, should be limited to crimes in the upper-middle seriousness-range. If imprisonment is a permissible option for lower-level crimes, the whole scheme of non-custodial sanctions is endangered.[16]★

Should there be any policy-based limits on substitutability? A desert

---

individual, does have added preventive impact—perhaps, by providing controls over the offender's activities during leisure hours. In practice, however, the impact may be very modest, if traceable at all. See J. Petersilia and S. Turner, 'An Evaluation of Intensive Probation in California' (1991).

★ If incarceration is a permissible option for low-ranking offences, and if non-custodial penalties are set at levels to achieve parity with this option, then the latter sanctions would have to be quite tough: say, substantial stints of community service. Middle-level crimes would have to receive still tougher sanctions, and it would not be easy to devise the requisite non-custodial measures: how many months of community service could possibly be feasible?

The Swedish sentence-reform statute seeks to avoid this pitfall by restricting the use of imprisonment to middle- and upper-level offences. The English law contains a similar requirement, limiting the prison sanction to 'serious' crimes (Criminal Justice Act 1991, s. 1(2)(a)). How effective the latter requirement will be remains to be seen, however. See Ashworth, *Sentencing and Criminal Justice*, ch. 9.

Several US state guidelines restrict 'imprisonment' (i.e. confinement in a state institution for a year or more) mainly to the more serious offences. For offences ranking lower on the penalty scale, however, the existing guidelines impose few limits on discretion and permit the judge to choose between probation and 'jail' (confinement in a county institution for less than a year). When extending the guideline system to these offences is discussed, however, there seems to be reluctance to abandon the jail option—even for the least serious crimes. If this option is retained, however, and if it is used as the basis of severity-equivalent interchanges with other

rationale could allow varying degrees of substitutability—from none to a wide degree. At one extreme would be a penalty scale with *no* substitutions permitted. The scale would consist of bands of penalties of ascending onerousness, with (say) cautions and small fines at the bottom, more significant financial penalties in the middle, attendance-centre requirements or home detention next, and full custody at the top. Within each band, a single type of sanction would be prescribed: for crimes in the upper-intermediate range of gravity, for example, substantial monetary penalties (relative to income) might be the only sanction prescribed. Such a scheme does grade conduct according to the seriousness of the offence, as proportionality requires. It could also significantly reduce the use of imprisonment, provided that custody is restricted to crimes that are serious. Nevertheless, the scheme's rigidity would probably make it unworkable. The prescribed sanction in a given band, for example, may not be capable of being carried out for certain offenders. Financial penalties for crimes in the upper-intermediate range, for example, could not be collected from offenders lacking in any regular earnings.

At the opposite pole would be a 'full-substitution' scheme. Any penalty may be substituted for any other, provided that the resulting severity is not altered. One might conceive of a sentencing grid which prescribes, not particular penalties, but 'sanction units' instead, that is, units of comparative severity.[17] A crime of upper-intermediate gravity, for example, might be punishable by a prescribed number of sanction units. These sanction units then would be translatable into actual penalties, depending on the latter's severity. The appropriate penalty-unit sentence could, then, be carried out through a variety of sentencing options. Different penalties could be combined, so as to achieve the prescribed total number of units.

Such a scheme, however, would have significant practical drawbacks. It presupposes a degree of sophistication in the ability to calibrate the onerousness of different sanctions that is not likely to exist. In a scheme with limited substitutions, one can make judgements that the prescribed sanctions are arrayed in ascending order of severity. The more numerous and easily combined the sanctions are, the more difficult and elusive the task of comparing severity becomes. Moreover, it is not clear why full

sanctions, those become unworkable—because the non-custodial sanctions would have to be as tough as the jail option.

What is to be done in those circumstances? The best solution would be to persuade the rule-maker to eliminate the jail option for low-ranking offences. If this cannot be accomplished—as it sometimes cannot for political reasons—then the next alternative would be to treat the imposition of jail as an upward departure from the proportionality requirements of the guidelines. Then, the (normally-applicable) non-custodial penalties for low-ranking offences can be set at the modest levels they should. If the rule-maker is unwilling to make even this concession, I think it still would be preferable to scale the non-custodial penalties beginning at the appropriate low severity levels. Then efforts can be made over time to persuade judges to invoke the jail option sparingly—with the aim of having that option treated eventually as a departure or eliminated entirely.

substitutability is needed. We have identified two kinds of situations where substitutions might be useful: (1) where the usual sanctions cannot feasibly be enforced and a more easily-enforceable substitute is called for, and (2) where a substitute might have some enhanced preventive effect, as in Professor Bottoms's previously-cited example. There may be some other situations as well, but it is not obvious what purpose would be served by cafeteria-style sentencing with unrestricted substitutions.*

What remains is *limited* substitutability: some sanctions preferred over others of equivalent severity. There are a variety of ways in which limited substitutability could be accomplished. One is suggested by the von Hirsch–Wasik–Greene scheme. One type of punishment would be pre-scribed in each range of sanction severity, and would be the normally-recommended type of disposition. However, the substitution of other specified sanctions (of comparable severity) could be invoked for certain reasons. These reasons could be preventive (where, for example, the sentencer had special reasons for believing that the alternative sanction might better induce the offender not to re-offend), or administrative (where, for example, the standard sentence could not effectively be enforced against this type of defendant).[18]

An alternative would be to prescribe several severity-equivalent sanctions in each band. Then, other criteria—including the preventive and adminis-trative one just mentioned—would be used to decide which sanction is preferable in the particular case.[19] The aim, however, is similar: to allow some additional flexibility without going to the extreme of a smorgasbord of penalties.

3. PENALTIES FOR BREACH

Non-custodial sanctions require back-up sanctions. Something needs to be done with the individual who, say, refuses to pay a financial penalty even though he is capable of paying.

Traditionally, the ultimate breach sanction was imprisonment. The form of traditional non-custodial sanctions made that clear. Probation, for example, is a conditional disposition whereby a sentence of imprisonment is put in suspension, provided that the offender abides by specified conditions of supervision in the community. If those conditions are violated, the prison sentence held in reserve may be invoked.

Relying on incarceration as the breach sanction is problematic, especially for the new intermediate sanctions. Offenders receiving these sanctions often have more extensive requirements to fulfil than those on traditional

* Note our objections to a full-substitution scheme relates thus to these above-stated practical difficulties. There is no objection in desert theory *per se* to such a scheme.

probation, and their compliance tends to be more thoroughly monitored. More frequent violations are thus likely to be uncovered. Easy resort to confinement for breach might well mean that *more* such offenders will end up in prison than would have been had those sanctions not existed at all.

When someone violates the conditions of his punishment, the breach can be seen as involving three distinct components. (1) The offender has not yet completed his original punishment, and still 'owes' the uncompleted portion in some comparable form. (2) The act of breach is, arguably, itself a reprehensible act, calling for some added punishment. (3) The breach, if more than a technical violation, may involve new criminal conduct. With these three elements identified, it is evident why a desert rationale would restrict the severity of breach sanctions; thus:

Element (1) may call for the uncompleted portion of the original penalty to be served in some different, more easily enforced form. It does not, by hypothesis, call for any additional severity.

Element (2) would call for a modest addition, depending how reprehensible the act of breach itself is deemed to be. The question is, how reprehensible? Victimizing acts can be gauged in their gravity by the extent to which they typically affect a victim's quality of life, * but the act of breach *per se* lacks a victim. No theoretical framework thus exists at present for judging the reprehensibleness of such conduct. Intuitively, however, breach seems hardly comparable with acts (say, of violence or deception) that typically should be deemed serious enough to warrant resort to incarceration.

Element (3) depends on the character of the new criminal behaviour. Imprisonment would thus be warranted on this ground, only if that behaviour would independently be deemed serious enough to warrant imprisonment.

On the basis of such arguments, the von Hirsch–Wasik–Greene scheme recommends only a modest step-up in severity: breach would result in the offender's receiving a sanction that is one band in severity up from the band in which his initial penalty was located.[20] Whether this particular rule should be adopted depends on the character of the penalty structure—and also on one's belief about the degree of reprehensibleness of element (2)— the act of breach. The principle, however, should be apparent: breach should occasion only a fairly modest increase in severity, and imprisonment should be very sparingly invoked as the breach sanction.

4. THE MORRIS–TONRY MODEL

So much, for the moment, for intermediate sanctions under a desert model.

---

* See Ch. 4, above.

What alternative approaches are there? The one alternative that has been put forward to date is Norval Morris's and Michael Tonry's.[21]

Morris and Tonry would rely, not on desert principles, but on 'limiting retributivism': desert would be seen as chiefly providing certain outer constraints, within which penalties would be set on other grounds.* The authors propose 'establishment of "exchange rates" to achieve, for appropriate cases, principled interchangeability between prison and non-prison sentences and among different non-prison sentences'.[22] Equivalence, they emphasize, does not mean equality, or even approximate equality, in the severity of sanctions involved. (That standard of equivalence, they say, would be suited to a desert-based scheme, which they reject.) Substitution would instead be permitted among penalties of significantly differing severities. The onerousness of the sanction thus would not define equivalence, but merely concern the outer desert constraints: that no penalty involved be plainly disproportionate in leniency or harshness to the gravity of the conduct.

Instead, Morris and Tonry speak of equivalence of *function*.[23] One penalty may be substituted for another if they both serve the same penological purpose in the circumstances. For a drug-using repetitive thief, for example, a sentence having some incapacitative effect might be desirable. The sanctions serving this function and hence functionally equivalent, they say, might either be a period of confinement, or house arrest with electronic monitoring and frequent unannounced drug testing. One penalty thus may be substituted for the other even if their penal bite is rather different.

The two crucial conceptions in Morris's and Tonry's scheme—those concerning desert limits and interchangeability of function—become elusive on closer examination, however. Consider, first, their view of desert. The authors do say that there should be certain desert-based constraints on the use of intermediate punishments. One proposed constraint is that inter-changeability should not be permitted for offences presumptively punished by imprisonment of 24 months or more.[24] The idea seems to be that such offences are too serious to warrant intermediate sanctions; and that for such offences, interchangeability would lead to too-great disparities.

A further desert-based constraint is offered, that the penalty must neither 'depreciate the severity of the crime and the criminal's prior record', nor 'impose pain or suffering in excess of that justified by the crime and by the record'.[25] No gloss, however, is supplied on the meaning of this vague standard. The authors assert that a precise formula cannot be supplied: the desert limits would have to vary with other features of the system, such as the breadth of the various offence categories, and the number of seriousness-gradations. What is troublesome about Morris's and Tonry's

---

* For discussion and critique of this view, see the previous Chapter; and, more fully, my *Past or Future Crimes*, chs. 4, 12.

formulation, however, is not that it lacks precision (what penal theory can be precise?), but that it provides *so* little guidance concerning the degree of emphasis that should be given to crime seriousness in deciding punishments. Thus:

1. One interpretation of the authors' model is that desert would substantially shape (albeit not fully determine) the gradation of penalties. This would approximate the 'range' models I discussed in the preceding chapter. Non-custodial sanctions might be graded into several bands according to their degree of punitiveness, ranging from the mild through the rather onerous. Substitutions would then be permitted within a given band (or at two bands' boundaries), but would otherwise be restricted. Thus a month's incarceration and a specified number of weeks of community service might be interchanged, as being both fairly onerous even if not equally severe; but the month in confinement could not be interchanged with moderate financial penalties. Some passages of their book suggest the authors support such an interpretation—for example, their citation with approval of the Minnesota sentencing guidelines, with their fairly narrow ranges of permitted punishments.[26]

2. Another interpretation is that desert should be invoked only to bar *manifest* disproportion, so that the scaling of non-custodial penalties would be decided chiefly on non-desert grounds. Passages of the book criticizing desert theory in rather scathing terms[27] could be cited in support of such an interpretation. On this view, substitutions of quite divergent penalties would be permissible: for example, imprisoning some offenders for up to a year, and subjecting others to a fine of, say, a month's earnings.

The 'equivalence of function' criterion also does not supply much assistance, in the absence of well-delineated desert constraints. That criterion does not necessarily concern justice at all, since the assumed functions or aims being compared may be those of crime control. Penalty X's 'functional equivalence' with penalty Y may consist in no more than its ability to incapacitate, or deter, with comparable or greater efficiency.

To decide whether two penalties are functionally equivalent, one would have to ascertain what the functions of the sentence are, and how well the different penalties can achieve those functions. The functions of the sentence are largely indeterminate under Morris's and Tonry's theory, however. The authors say, for example, that where the governing aims of the sentence are 'retributive and deterrent', a short stint of incarceration and a substantial fine might be functionally equivalent.[28] How does one tell, however, when these are the appropriate purposes? The reference to retributive purposes is puzzling, because the authors simultaneously assert that retribution (i.e. desert) provides only limits and little or no guidance in deciding the particular sentence. The deterrent purpose is equally uncertain, because no suggestion is made about deciding when this aim—rather than some other—is the desired purpose. The judge, they assert, should determine the aims of the sentence in the particular case.[29] Alas, if he or she is uncertain on the choice of aims, no guidance is provided.

How well would Morris's and Tonry's scheme fare in practice? Let us try it out on the three problems that chiefly have been troubling intermediate-sanction policy. The first is that of using intermediate sanctions for lesser crimes. The sponsors of 'alternatives to incarceration' are understandably concerned about the public credibility of their programmes, and credibility is enhanced through high rates of completion and low rates of breach of programme conditions. This prompts recruitment from the 'shallow end' of the offender pool, that is, use of more tractable offenders as participants, who tend to have modest crimes and criminal records. (This has been occurring with several much-touted intensive-supervision programmes in the USA, for example.[30]) The result is merely to increase punishment levels for the least serious cases.

How is recruitment from the shallow end to be prevented? On a desert-based scheme, as we saw earlier in this chapter, penalties of intermediate severity would be barred in principle for offenders convicted of low-ranking crimes. How about the Morris–Tonry scheme? It is difficult to tell, given the uncertain nature of its desert constraints. The wider the limits are, the less serious could be the crimes of those who receive the intermediate penalties.

A second problem is sanction stacking: imposing multiple sanctions on an offender (such as a fine, supervision, *and* something else). Sanction stacking makes the severity more difficult to gauge. It also makes compliance more complicated and thereby increases the likelihood of breach: the more different kinds of things the offender is called upon to do, the less likely it is that he or she will do them all. The individual whose transgression stems from difficulty in abiding by the normal rules of civil existence is not likely to abide by a plethora of new penal requirements.

The von Hirsch–Wasik–Greene proposal would largely bar sanction stacking through its restrictions on proliferation of sanctions. Morris and Tonry offer no such restrictions. Provided the loose interchangeability criteria are met, the judge would be free to choose any of a variety of intermediate penalties, to invent new variants of his own,* and to pile several such sanctions on particular offenders.

In making such choices, the judge would be free to determine what purpose the particular sanction is designed to serve. Giving judges wide leeway to choose 'the purpose at sentencing'[31] raises the question of how unsupported judgements are to be kept in check. Consider the judge who harbours the supposition that 'short sharp shocks' have a marvellous

---

* For the problems of 'bespoke' penalties, designed *ad hoc* for the particular occasion, see Ch. 8, below.

deterrent effect. How free should he be to adopt deterrence as his purpose at sentencing, and invoke stints of jail as the preferred deterrent?

Most worrisome, still, are breach penalties. A desert rationale—for reasons discussed already—would sharply restrict the use of imprisonment as the breach sanction. Under the Morris–Tonry model such restrictions weaken, because the desert constraints are so much more uncertain. Indeed, the authors do not explicitly address breach at all.

Would Morris's and Tonry's 'equivalence of function' criterion help? Not much. A defendant who violates the terms of his non-custodial punishment may be perceived as presenting a higher degree of risk. Imposing greater restrictions may thus seem warranted on incapacitative grounds. Severe breach sanctions, such as imprisonment, may be seen as 'functionally' equivalent—as an effort to pursue the crime-prevention aims of the sanction by more restrictive means, the initial milder response having failed.

5. CAN FULLER GUIDANCE BE PROVIDED, WITH MIXED PENAL AIMS?

The question raised by Morris's and Tonry's book, but not satisfactorily answered, is whether one can provide meaningful guidance on non-custodial penalties, while adopting a mixed set of sentencing aims. This is possible, but only through more stringent constraints on the choice of sanctions than those authors have been prepared to suggest; thus:

1. Penal equivalence should mean approximate equivalence in penal bite, that is, in severity. Penalties that are significantly different in onerousness are simply not 'equivalent', in any ordinarily understood sense of the term. On a desert rationale, substitution among penalties of differing severities would not be permitted; on mixed rationales, it would be to a greater or lesser extent. Even on the latter kind of theory, however, an important moral difference exists between (*a*) replacing a sanction with another comparably severe one, and (*b*) replacing it with one significantly differing in severity. The former stands less in need of special justification, because neither the extent of the offender's suffering nor the extent of the implicit censure alters. That difference should not be obliterated by speaking of divergent penalties as 'equivalent'.

2. To the extent that substitution among penalties of differing onerousness is permitted, there should be *meaningful* limits on those permissible differences. While hybrid models of the kind discussed in the previous chapter have somewhat less stringent proportionality constraints than a desert model, proportionality remains the primary determinant for the comparative severity of punishments. Vague formulations of the appropriate desert limits, such as Morris and Tonry propose, will not help.

3. There should be certain other, policy-based, limitations. Proliferation

and stacking of sanctions should be restricted. Even within applicable constraints on severity, moreover, there should be some guidance concerning the kinds of reasons the judge may rely upon in choosing the sanction. Rather than giving the judge unfettered freedom to decide the 'purpose at sentencing' which Morris and Tonry suggest, there should be procedures for scrutinizing the rationality of that purpose and its achievability in the situation at hand.

Mixed schemes that meet these three conditions would be tolerable—in the sense of avoiding recruitment at the shallow end, sanction stacking, or routine imprisonment for breaches. The question remains, however, of what positive advantages such mixed schemes could offer. The relaxation of desert constraints would make interchanges somewhat easier: sanctions having differing 'bite' could be replaced for one another, provided the difference is not large. Some—albeit not much—added severity would also be permitted for the back-up sanctions. But what could such added flexibility achieve?

If the aim is to achieve added crime-prevention, I am sceptical. Crime rates simply do not seem sensitive enough to modest variations in comparative penalty severity. If the aim is to achieve added 'parsimony', I am again doubtful: limited deviations from ordinal desert are unlikely to affect penalty levels much. There also remains the difficulty, noted already, of ensuring that the added flexibility does not support penalty *increases* for a significant number of offenders.★ If parsimony is the aim, the scenarios of how it is to be achieved would need to be set forth carefully.

A final aim might be to simplify the system's administration. One could substitute among penalties without having to be quite so exacting about their comparative severity; and one could be more flexible in devising back-up sanctions. While that is a practical advantage, beware the pitfalls: we may not want much added severity for the back-up sanctions, for example, lest we find ourselves back to imprisoning violators routinely. Again, carefully-drawn scenarios would be needed—to see exactly how the added flexibility would work in practice, and what its practical advantages and hazards are. General claims that mixed schemes are simpler or easier to operate do not suffice.

A final warning: increasing the use of intermediate sanctions is an attractive social experiment. These could become the penalty of choice for crimes of middling seriousness—those which now too often are responded to by inappropriate extremes of imprisonment on one hand, or perfunctory responses on the other. Scaling principles, such as those suggested in this chapter, should help assure that the penalties are used in this fashion. The experiment, however, has its perils. Without adequate controls on their use, intermediate sanctions could routinely become invoked for relatively minor

---

★ See discussion at end of Ch. 6.

criminal conduct. Such sanctions could also become the vehicle for too easy subsequent incarceration as a penalty for breach. If we are not careful about who gets these penalties, for what kind of conduct, and subject to what disciplinary rules, intermediate sanctions could become yet another route to escalated sanctions and and a widening of penal controls.

8

Penance and Personalized Desert

THE personalized sentence, designed to fit the defendant's personality and needs, was associated with penal rehabilitationism. Faith in it has waned for a variety of reasons. Treatment no longer seems plausible as the chief aim for sentencing,[1] and the wide discretion required for such an aim has turned out to lead, too easily, to inconsistency and caprice.[2] Conceivably, however, the personalized sentence could be reconceptualized to achieve another aim—that of conveying blame particularly suited to the wrong-doer. When someone is blamed for a wrongful act, shame or penitence would seem to be the appropriate response. Conceivably, the sentence might be fashioned so as best to elicit that response. Controls over excessive discretion could be introduced, to prevent disproportionate and inconsistent outcomes. Would such an approach make sense? I think not, for reasons I shall explain.

1. 'SHAMING'

John Braithwaite puts forward the idea of 'reintegrative shaming' in a volume[3] published just before his and Pettit's book, discussed in Chapter 3. The sentence, Braithwaite asserts, should convey an appropriate 'shaming' message to the offender. The form of the message should depend on the individual's receptivity; its function should be to reinforce his or her inhibitions against offending. Gestures of re-acceptance should be provided in the event the offender responds; indeed, the message contained in the sanction should make this potential for re-acceptance clear. To accomplish these ends, the sentence should be suited to the individual case. While proportionality of sentence may suffer as a result, that is is of little concern to him (as we have seen already).

Braithwaite's proposal amounts to a quasi-rehabilitative strategy: a therapy for reducing the offender's inclination to offend again. The novelty lies merely in the therapy technique: that is to consist of shaming rituals, instead of psychological counselling or job training. How much shame to elicit, through what sanctions, would depend on what will most likely make the offender desist in future. He does not see the censuring aspect of the sentence as conveying disapproval that should fairly reflect the degree of reprehensibleness of the defendant's conduct. Braithwaite's notion of 'shaming', in other words, has little to do with the conception of censure I have sketched in this volume.*

2. PENANCES

R. A. Duff would also favour a considerable degree of individualization of sentence,[4] but his reasons are of greater interest—for his is a genuine censure-based desert theory. Let me, then, highlight the differences between Duff's view of censure and mine; and how those differences lead to divergent conclusions on individualization of sentence.

My view of the censure conveyed through punishment, outlined in Chapter 2, deals with the person *externally*. The disapproval conveyed by the sanction gives the actor the opportunity to reconsider his actions and to feel shame or regret. However, it is left to him to respond: censure, on my analysis, need not specially be fashioned to elicit certain sentiments in him— whether those be shame, repentance or whatever. That is why, I argued, there is no need to try to suit the censuring response to the actor's degree of receptivity.

Duff's penance view, however, is meant to elicit certain *internal* states. The disapproval conveyed by the punishment and its accompanying hard treatment, he argues, helps 'induce or to reinforce repentance, by forcing [the actor's] attention onto her wrong-doing (which she might be otherwise unwilling to face) and by providing a structure within which she can attend to what she has done'.[5] The imposition of the penance thus aims at inducing a penitent understanding. However, Duff is not advocating a purely consequentialist view such as Braithwaite's. What matters is not just the actor's being induced to feel shame and thus, perhaps, restrain himself

---

* For the basic objections to Braithwaite's conception of shaming, and its difference from the view of censure advocated in the present volume, see Ch. 3 above.

Even judged in its own terms, however, Braithwaite's notion of 'reintegrative shaming' presents a host of difficulties. First, how can one ensure that shame is elicited? The sanction can convey blame or disapproval, but *shame* is a responsive attitude in the offender (see Ch. 3, above). To the extent that offenders are disaffected from the wider community, official disapproval may bring forth no shame. Second, what is the link between shame and desistence? If other incentives for criminality continue to exist, the ashamed offender may continue offending anyway. Third, how can the reintegrative element (that is, Braithwaite's suggested gestures of re-acceptance) have any meaning? *Real* reintegration is the most ambitious version of rehabilitative success, and is notoriously difficult to achieve: it involves not just inducing the offender to cease from offending, but creating or reviving in him a sense of solidarity with the community. This success is scarcely achievable by having the sanction express an intent or desire to re-welcome the responsive offender. (Such expression may amount to little more than a new official piety—a 'We Officially Still Love You Anyway'—reminiscent of the pieties of traditional juvenile justice.) The actual effectiveness of Braithwaite's strategy in inducing offenders to desist is, of course, a question for empirical inquiry. Experimentation with his technique, I understand, is being undertaken with groups of juveniles in a few localities. But the strategy does not seem to me to hold much promise on its face, and I would be surprised if successes are achieved—other than, perhaps, with narrow subgroups of amenable offenders.

better—but his reasons why. The censure communicated by punishment, according to Duff, is supposed to *convince* the actor that his act was wrong; his sense of shame, and subsequent efforts at desistence, ought to stem from that conviction. Mere embarrassment at being shamed, or fear of ostracism would not be enough—even if it did lead to better future behaviour. It is because the censure is supposed to give the actor a *moral* reason for responding, that the degree of blame conveyed should be consistent with blameworthiness of the conduct. Duff, unlike Braithwaite, thus does subscribe to the principle of proportionate sanctions.[6]

Which is preferable, my 'external' view, or Duff's 'internal' one? In Chapter 2, I argued that my own view comports better with the logic of censure: that, when censuring, we act more as judges than as abbots. The fact that we are prepared to censure the already-penitent and the seemingly incorrigible, the fact that we ordinarily do not inquire into the actor's receptivity when censuring, all point to such an externalist view of censure.

Duff, however, could concede this last point—that ordinarily, censuring differs from imposing a penance—and still insist on his theory. He would simply argue that punishment, properly conceived, should do more than express mere censure: that it *should* serve as a penance. This is the claim that needs to be answered.

Duff draws his illustrations from contexts where his penance view makes sense. When the monk misconducts himself, it is quite appropriate for his superior to impose a sanction designed expressly as a penance, for the latter's duty is to promote his charges' spiritual development. Merely dealing with the wrongdoing externally—that is, imposing a condemnatory sanction that gives the monk grounds for feeling shame—would be insufficient: it would not meet the abbot's responsibility regarding the erring monk's moral welfare. And respect is shown for the monk's moral agency, in that the sanction is designed to bring that person *himself* to a penitent understanding of the wrongfulness of his conduct.

Legal punishment, however, involves a different setting. The State cannot be expected to have the kind of insight and sympathetic under-standing that an abbott should have for a small number of voluntarily-enlisted charges explicitly sharing a common purpose. And citizens have not entrusted their moral development and spiritual welfare to the State. Why, therefore, should the State be entitled to use its coercive powers to seek to induce moral sentiments of repentance? Duff's view seems to involve a strong moral paternalism. True, the conduct at issue involves harm to others rather than to the self. Duff's focus, however, is on the quality of the actor's own moral response to his harmful act.*

* In Joel Feinberg's terminology, this would not be paternalism *simpliciter*—as the aim is not to prevent actual harm to the actor himself. But it is what he calls *moral* paternalism—the aim being to foster some desirable moral state (in this case, repentance) in the actor. See J. Feinberg, *Harm to Others* (1984), p. xiii.

A supporter of Duff's view might concede that the State should not be entrusted with the broad authority of an abbot over the general moral development of its citizens. The State, he or she might argue, should concern itself in the criminal law primarily with a restricted subclass of duties that relate to not injuring other persons, and to observing certain basic obligations of citizenship. Where these duties are involved, however, why should the State not visit penances as Duff has suggested?

To answer that question, it is worth noting that Duff's examples assume that the person punished acknowledges the moral authority of the punishing agent. In the context of the criminal sanction, that assumption becomes problematic. Consider the disaffected defendant who (for reasons of radical opinion, previous bad experience with the criminal-justice authorities, or whatever), does not think that the State possesses any particular moral authority over him. Whatever regrets he may have about his action, he does not wish to respond to any demand by the *State* that he feel penitent. Suppose that State's sanction is imposed upon this person as a penance. A penance seeks to reach deeper than mere penal censure does: in order to elicit the requisite attitudes of repentance, the sanctioner needs to inquire into the person's feelings—or at least, fashion the sanction so that it is designed to reach those feelings. Might this not be overreaching on the State's part? Granted, it need not be manipulative literally: Duff's penance is designed only to help the person reach his *own* penitent understanding, not to compel him to express sentiments he does not feel.★ Might not the person object, nevertheless, that the penance is an inappropriate form of State intrusion, that while he should be treated as *capable* of moral judgement, the nature of his *actual* attitudinal response to the State's censure is his own business? For the monk, no comparable problem of overreaching exists. By taking orders, he has undertaken to be guided by certain moral precepts and has acknowledged the spiritual authority of the abbot. Even if the particular penance is imposed against his wishes, it is part of a regime whose purposes he has accepted. If he thinks that religious superiors lack standing to seek to elicit penitence from their charges, he should choose another vocation.★★

Answering this objection would require a particular, and fairly ambitious, account of State power—one in which the State not only expresses its disapprobation of certain conduct, but tries to bring about certain responsive attitudes in those whom it condemns. Perhaps such an account can be defended—although I would not be easily persuaded. But the account could not be derived simply from the notion of treating the offender as an agent capable of moral judgement.

★ For further objections to compulsory attitudinizing, see the next chapter.

★★ For an elaboration of these themes, see U. Narayan, 'Moral Education and Criminal Punishment' (1993).

In earlier writings, Duff asserts that his penance model is offered only as an ideal for the penal sanction: most legal systems may not have sufficient time, sympathy, or insight to make a system of penances workable in practice.[7] However, what is at issue is whether his view of legal punishment should serve even as an ideal: whether the State (even under the best of circumstances) should strive toward an abbot-like role of seeking penitence from erring citizens.

Recently, moreover, Duff has indicated that his view is not just an ideal; that it may be possible, to some degree, to make some punishments (particularly non–custodial ones) function more like penances.[8] Even partial implementation, however, would raise a number of difficulties. Thus:

1. *Questions of Feasibility*. The moral functions of censure which I sketch in Chapter 2 seem modest enough to be feasible. A condemnatory sanction does appeal to the sense of right of ordinary citizens, and thus provides them with a reason for not offending. Actual offenders are visited with punishments that treat them as moral agents, not tigers. Duff's conception, however, is more ambitious. Penances deal with actual offenders only, and assume that such persons, or a considerable number of them, will respond with feelings of penitence. In a monastery, such an assumption seems realistic: most monks are likely to take seriously the penances visited on them. But in the rough-and-tumble of the penal system, how realistic is it? Cases where punishment functions as Duff describes—as a penance leading to actual penitence—may be the exception rather than the rule. If exceptional, should a whole penal strategy be built around them?

2. *Problems with Maintaining Proportionality*. My suggested view of punishment readily allows maintenance of proportionate sanctions. The severity of the sanction expresses how reprehensible the act is deemed to be. What responsive attitudes the punished actor adopts—whether or not he reacts with shame or penitence—is left to him. If A and B commit comparably reprehensible acts, they should receive sanctions that signify the same amount of disapproval. If those sanctions are enough to induce A to be penitent but not B, so be it.

Duff acknowledges the importance of proportionality, and states that his theory is preferable to conventional rehabilitationism in that it would not 'justify continuing a criminal's punishment until she is reformed'.[9] Proportionality needs to be maintained, he asserts, to ensure that the punishment communicates a condemnation appropriate to the gravity of the crime.

Nevertheless, Duff's penance rationale creates an undesirable tension with proportionality. The latter principle calls for A and B to receive comparably severe punishments, if the gravity of their crimes is approximately the same. The aim of eliciting penitence, however, points the other way: B might have a thicker skin than A, and hence might need a tougher penance before the message is likely to penetrate. One could, of course, propose that proportionality governs the severity of the punishment and penance its mode: so A and B would receive penalties that have to be comparable in severity, but might differ in other respects so as better to elicit penitence.★ This, however, just illustrates the tension: such restrictions on severity would make it more difficult to suit the penalty to the actor's expected responsiveness; on the other hand, easing those restrictions in order to facilitate the achievement of penitence sacrifices proportionality.

3. *Mikado-Style Sanctions.* Duff relies upon his penance theory to promote the Mikado's view of punishments: the penalty should illustrate the character of the crime, in order to promote reflection on its evil. Crimes motivated by greed should thus receive fines, violent crimes (because destructive of the community's basic values) should be punished by a period of exclusion from the community (i.e. imprisonment), and so forth.[10] But the Mikado principle, if taken seriously, will lead to a proliferation of modes of punishment, as ingenuity is employed to suit the manner of the punishment to its crime. Proliferation makes severity-comparisons among punishments harder, as noted earlier.[11]★★

4. *Bespoke Sentences.* Duff endorses, as a vehicle for penances, the 'bespoke' sentence: that is, the penalty specially devised for a particular defendant's case. He speaks with approval of an American case in which a convicted slum landlord was sentenced to live in his slum building for a specified period. This, Duff asserts, is an 'apt way both of confronting him with the nature and effects of his crime, and of giving him the opportunity to restore and repair his relations with his tenants'.[12] The bespoke sentence, however, has manifold drawbacks. Maintaining proportionality requires an ability to compare sanctions' severity. The more idiosyncratic the sentence, the more difficult the comparison becomes. How does one compare the onerousness of being required to live for fourteen days in a tenement flat with another bespoke sentence—say, having to give a specified sum to charity? Tailor-made sentences, like tailor-made clothes, tend also to be reserved for the affluent. The slum landlord can be sentenced to live in one of his tenements, but the ordinary burglar cannot because he may live there already. Enforcement is also a problem. With generally-available sanctions, regular procedures can be worked out to help ensure compliance: e.g. to help insure payment of day-fines or attendance at community-service work sites. Evasions can be anticipated with experience, as they cannot be with individually-devised sentences. The slum landlord, in the American case, was able to thwart the intent of his punishment by bringing along his own repairmen and guards.[13]

Duff might concede the foregoing practical difficulties, but argue that the

★ For discussion of severity-equivalence, see Ch. 7 above.
★★ See Ch. 7 above. Inviting sanctioners to give defendants a 'taste of their own medicine' can also too easily lead to degrading or intrusive penal routines. For discussion of this latter issue, see Ch. 9.

aim of inducing penitence is worth risking them. But is it? I have suggested above that it is at best debatable whether the State *should* seek an abbot-like role in dealing with criminal conduct, and that it is very unlikely to be able to perform such a role effectively. I doubt that we should risk compromising proportionality among penalties to achieve the theoretically debatable and practically hard-to-achieve end of making criminal sanctions function as penances. The punishment should convey blame; and blame gives the offender reason for an appropriate moral response. But it should not be the business of the State to try to engineer that response—and there is no need to personalize the sentence in order to do so.

3. PROPORTIONALITY AND STANDARDIZATION

If proportionality is given the pre-eminence I have been urging, sanctions should vary in their severity with the seriousness of crimes. Ratings of crime seriousness call for a certain measure of standardization. Seriousness ratings should apply in the first instance to standardized categories or subcategories. The categories or subcategories need not necessarily be broad: there could be considerable differentiation beyond the statutory offence descriptions. Departures from the norm should be permissible for significantly deviant special cases. To deviate, however, the case should involve more than a certain threshold quantum of difference in gravity from the standard case.

The most obvious reason for having a modicum of standardization concerns consistency of judgement. Different sentencers should apply similar norms of seriousness in their judgements, and this is facilitated by assigning seriousness-ratings to offence categories or subcategories. Un-usual cases can then be accommodated through deviations from these ratings.[14]

There are, however, two further reasons for preferring a degree of standardization in judgements of seriousness. One concerns the concept of criminal harm, and has been mentioned already.* A living-standard conception of injury itself involves a degree of standardization: the living-standard refers, as we saw, to the *standardized means* for achieving a certain quality of life.

The second reason concerns the culpability. What is at issue in deciding punishments is how blameworthy the *actor* is in committing the conduct, not how the victim should be compensated for his or her injury. This actor perspective requires consideration of how much, and within what limits, the act's harmful consequences can justly be attributed to the actor. The issue, here, is not only one of the forseeability of the consequences but of an

* See Ch. 4.

actor's opportunity to *appreciate* them. The offender who is convicted of a given offence can be expected to appreciate its normal and predictable consequences, and that is why it is appropriate ordinarily to hold him to account for those consequences in assessing the gravity of the conduct.[15] The more unusual the consequences are, the harder it may be for him to comprehend and empathize with their impact on the victim's life.

If a degree of standardization of desert judgements is called for, how should it be achieved? The techniques vary, and none will be perfect. Having discussed them at length in previous writings,[16] I shall only summarize my views here.

One possible device is numerical guidelines. Categories or subcategories of offences are assigned seriousness-ratings. A numerical grid is then provided, containing a series of bands of ascending severity. Imprisonment would be the sanction in the highest band, with various types of non-custodial sanctions prescribed for the remaining bands—plus some measure of limited substitution. Decision-makers may deviate from these prescriptions where aggravating or mitigating circumstances are present. Minnesota and Oregon have such schemes operative for the use of imprisonment[17]— but a full grid, in which both imprisonment and non-custodial sentences are dealt with, has yet to be adopted.

The alternative is to use statutory sentencing principles, which have been adopted in Finland, Sweden, and (since 1991) England.[18] All three countries make proportionality the main governing standard, for the imposition both of imprisonment and of non-custodial sanctions. The appellate courts are expected to apply that standard in developing a tariff of penalties.

Which approach, numerical standards or statutory sentencing principles, is preferable? Both can be done well or badly,* and either can reflect various possible rationales. Numerical guidelines are simpler for sentencing courts to apply and for appellate courts to enforce. However, a grid (even with allowance for departures for aggravation/mitigation) tends to be quite aggregative, lumping together behaviours of heterogeneous gravity. It is also vulnerable to political pressures: when the rule-making agency writes numbers on a grid, it must justify those numbers to the legislature and to criminal-justice constituencies.

Statutory sentencing principles are less aggregative, as the appellate courts remain free to continue drawing distinctions within crime categories among cases of differing gravity. They contain no numbers that legislatures would be tempted to raise during election years. They also might encourage the judiciary to think more systematically about proportionality issues, as judges would be applying that general standard in deciding cases. However, this approach depends crucially on the judiciary's willingness to implement

* The US federal guidelines are an example of numerical guidelines done badly, with excessive severity and without any stated underlying rationale, see Ch. 10.

the general statutory standards through case law, in a way that yields a workable tariff of penalties. To what extent the English and Swedish judiciaries respond in this fashion to these countries' recently-adopted statutory standards remains to be seen.

What form the guidance should take—whether numerical guidelines, statutory principles, or something else—may well vary with the jurisdiction involved. The choice depends on the jurisdiction's legal and political traditions, and on how much responsiblity the appellate courts are capable of undertaking. Numerical guidelines might thus be suitable for one jurisdiction, statutory principles for another. What needs to be determined is what form of guidance, in the particular jurisdiction involved, would hold most promise for providing a reasonable degree of standardization, with flexibility still remaining to deal with unusual cases.

9

Degradingness and Intrusiveness

Co-author: *Uma Narayan*

AMONG the less attractive forms of human ingenuity has been the capacity to devise degrading rituals for those punished. The history of imprisonment has been full of such rituals—ranging from the lockstep and the treadmill in early days, to aversive therapies in recent times. Prison-reform efforts have been devoted in considerable part to getting rid of such practices. Intermediate sanctions (despite their positive potentials, discussed earlier) can offer new opportunities for creating humiliating routines. The fact that such sanctions are relatively novel, and little controlled, makes this danger all the greater. We need to worry about how to rule out modern versions of the stocks.

The strictures suggested in the preceding chapter may be helpful, to some degree. If the sentencing judge is not to aim at imposing a penance, and if 'bespoke' sentences are rejected, he may have less incentive to use his ingenuity to give the defendant a taste of his own medicine. Those strictures, however, are not enough. Even standardized penalties can visit unjustified humiliation. We require a theory on what types of penalties should be ruled out because of their degrading or intrusive character.

1. COMMON FALLACIES

When we consider this issue, we enter little-explored territory. Whereas an extensive literature on proportionality exists, not much theory has been developed on what makes punishments unacceptably humiliating. We might begin by clearing away the undergrowth, that is, putting aside some commonly-heard fallacies.

One fallacy is the *anything-but-prison* theory. Intervention in the community is tolerable irrespective of its degrading or intrusive character, this view asserts, as long as the resulting sanction is less onerous than imprisonment. Because imprisonment (at least for protracted periods) is more severe than almost any community punishment, and because many prisons still resort to their own demeaning routines, it would seldom be possible to object.

The anything-but-prison theory is a version of the wider misconception that an individual cannot complain about how he is being punished if something still nastier could have befallen him. The idea bedevilled prison policy for years: prisoners should not complain of conditions in institutions because they might have fared worse, i.e. been held longer or in more brutal conditions, or even been executed. The short answer is that a penal measure needs to be justified in its own right, not merely by comparison with another, possibly worse measure.

The theory also assumes that there is only one dimension along which punishments may be evaluated ethically—that of their severity. If a non-custodial penalty is less severe than imprisonment, how can it be problematic? The assumption is mistaken. Severity is *a* dimension along which penalties are to be evaluated, and it is that which is governed by the principle of proportionality. Degradingness deals with another dimension. The stocks may, indeed, be less severe than incarceration; and thus, if the person involved has been convicted of a lesser felony, may be less violative of proportionality than prolonged imprisonment. The sanction is objectionable nevertheless, on the ulterior grounds of its extraordinarily humiliating character.

A second fallacy is that *intrusiveness is a matter of technology*. The installation of an electronic monitor on an offender's telephone elicits comparisons to 'Big Brother', but no similar issues of privacy are assumed to arise from home visits by enforcement agents. The mistake should be obvious: Orwell's totalitarian State may have relied on two-way television screens, but the Tsarist secret police achieved plenty of intrusion without newfangled gadgetry. The same point holds for non-custodial sanctions. Intrusion depends not on technology, but on the extent to which the practice affects the dignity and privacy of those intruded upon.

A third (and particularly American) kind of mistake is *legalism*. Degradingness, on this view, is a matter of whether the practice infringes on specific constitutional requirements. The US Constitution does not give much consideration to the treatment of convicted offenders, and such provisions as are germane have recently been reinterpreted by the federal courts so as to give scant protection. These provisions do not exhaust the ethical requirements that a State should abide by in its treatment of convicted offenders. This is now understood where proportionality is concerned: whereas the Eighth Amendment to the US Constitution (as now construed) scarcely may outlaw even the most grossly disproportionate punishments,[1] a fair system of punishment should observe more stringent proportionality requirements. The same point should hold for the present issue of degrading punishments. When a programme is developed, its sponsors should ask themselves not only whether it passes constitutional muster, but whether there are any substantial ethical grounds for considering it is unacceptably humiliating to the offender or intrusive on the rights of third parties.

2. THE RATIONALE FOR 'DIGNITY' IN PUNISHMENT

The basic objection to degrading punishments is that they fail to treat the offender with the dignity due to a human being. Jeffrie Murphy conveys this point well, in the following passage from a 1979 essay.

A punishment will be unjust (and hence banned on principle) if it is of such a nature as to be degrading or dehumanizing (inconsistent with human dignity). The value of justice, rights and desert make sense, after all, only on the assumption that we are dealing with creatures that are autonomous, responsible, and deserving of the special kind of treatment due that status . . . A theory of just punishment, then, must keep this special status of persons and the respect it deserves at the center of attention.[2]

What Murphy is saying is that convicted offenders are still members of the moral community: they remain *persons* and should be treated as such. The evil of degrading punishments is that they deny those punished that status.

Someone's status as a person would ordinarily militate against *any* sort of demeaning treatment. With offenders, however, there seems to be a complication—the nature of punishment itself. Punishment not only serves as a disincentive to certain conduct but conveys censure. Blame, because it embodies disapprobation of the offender for his or her conduct, is surely humbling. What is left, then, of the idea that punishment should not demean its recipients?

The answer lies in the difference between censure and attempts to humiliate. Censure may make someone ashamed of what he has done, for shame is a sentiment a moral agent may (indeed, ought) to feel when the wrongfulness of his behaviour has been brought to his attention. Degradation is something quite different: it is being treated as something less than a person, and being made to feel so. Indeed, degrading treatment interferes with any legitimate moral response the censured person may make. The more the offender is treated in a degrading fashion, the more he will feel demeaned simply by what is being done to him. When prisoners are made to walk the lockstep, shuffling forward with eyes averted, or when convicts are put in ridiculous postures as part of a community punishment, they are humiliated irrespective of any judgement they themselves are prompted to make about the propriety of their conduct. The feeling comes not from acceptance of the social judgement of censure, but simply from the fact that they are being treated as inferior beings.

Punishments, therefore, should be of a kind that can be endured with self-possession by persons of reasonable fortitude. These individuals should be able to undergo the penalty (unpleasant as it inevitably is) with dignity, acknowledging their guilt if they feel guilty, or not acknowledging it if they do not feel (or wish to admit) guilt. A person can endure the deprivation of various goods and liberties with dignity, but it is hard to be dignified while having to carry out abasing rituals, whether the lockstep, the stocks, or newer rituals.

In what ways may penal measures be objectionable on this ground? Three ways come to mind—although others may also exist. The most clearly objectionable measures are those that are *dehumanizing*, namely, actually or potentially destructive of the personality of the offender. These are the interventions that destroy (or substantially interfere with) a convicted person's ability to behave as a human being capable of feeling, reflection, and choice; by inducing states of extreme terror, depression, and the like. Torture, cited by Murphy,[3] is the paradigm. Another instance is long-term solitary confinement, with its depersonalizing effects. So would be various forms of brainwashing. These are most readily inflicted in a prison, but some can be carried out outside the walls, for example, supervision involving intense and prolonged forms of harassment.

Thus far, there should be little disagreement: who would admit they want penal measures designed to destroy or endanger the personality of an offender? There is, however, a less drastic kind of intrusion which should also be ruled out, namely, that of *demeaning rituals*. The stocks are the classic historical instance, but so would any other penal measure which requires the adoption of ridiculous or grovelling postures or attitudes. An example is 'boot camp' routines (now common in the USA as a punishment for young offenders), which involve submission to verbal abuse by staff and like rituals as part of a programme of brief 'shock' incarceration.[4] Such measures may not necessarily involve actual destruction of or threat to the offender's personality. One can survive boot camp with one's capacity for feeling, reflection, and choice intact. A person, however, is entitled not only to have his personality preserved from destruction, but *to be able to present himself to the world as an intact human being*. Demeaning rituals interfere with this latter interest: the person is required to adopt the posture of a fool, slave, or other inferior being.*

Another form of objectionable penal routine involves *compulsory attitudinizing*. An example is some American judges' practice of requiring convicted drunk drivers to attach self-accusing bumper-stickers to their vehicles. What is wrong with compulsory attitudinizing? While punishment conveys disapproval of the actor's conduct, any contrition or self-criticism expressed by the actor must—if his moral agency is to be respected—reflect his own views. If those views are inconsistent with the attitudes he is required to

---

* One can argue that such routines are doubly degrading: first, because the person presents himself in a ridiculous posture, and second because he is *made* to do so. Another problematical feature of demeaning rituals is that they often enlist the public: the routine is designed to elicit ridicule of the offender. This not only can produce excessive public responses (especially against offenders against whom there is prejudice); but also raises questions of *standing* to punish. There are good reasons—relating to predictability and impartiality—why the public is not the appropriate penal agent in a *Rechtstaat*. If the public does not have standing, it is all the more humiliating to suffer the punitive treatment at its hands.

express, then his agency is not respected. There is no way a person can, with dignity, be compelled against his will to admit himself to be a moral pariah.

3. ACCEPTABLE PENAL CONTENT

How should we apply these ideas? One way would be to try to identify and list the various kinds of measures we wish to rule out as degrading. But as violation of a person's dignity is a matter of degree, and as substantial variety of non-custodial measures may be involved, this would be no easy task.

A better approach, we think, is through the idea of *acceptable penal content*. The penal content of a sanction consists of those deprivations imposed in order to achieve its punitive and preventive⋆ ends. Acceptable penal content, then, is the idea that a sanction should be devised so that its intended penal deprivations can be administered in a manner that is *clearly* consistent with the offender's dignity, that is, manifestly not objectionable on one of the grounds just sketched. If the penal deprivation includes a given imposition, X, then one must ask whether that can be undergone by offenders in a reasonably self-possessed fashion. Unless one is confident that it can, it should not be part of the sanction.

Where prisons are concerned, we already have the kernel of this idea, expressed in the maxim that imprisonment should be imposed *as* punishment but not *for* punishment. The idea is that deprivation of freedom of movement should be the main intended penal deprivation—that while it is severe (and hence suitable only for serious crimes), such deprivation *per se* can be endured with a degree of self-possession. According to this maxim, the intended penal content should not include various ancillary routines within the prison which raise potential problems of being dehumanizing or degrading. It would be inappropriate, for example, to prescribe solitary confinement as the punishment for designated crimes. Notice, also, that it should not be necessary to determine whether each possible sanction-within-the prison is unduly degrading. That prison should serve only as and not for punishment functions as a prophylactic principle, to endorse only that deprivation—of liberty—which we think can decently be imposed, and not to authorize all kinds of further impositions whose moral acceptability is in doubt. Granted, the reality of US (and some English) prisons is different, with numerous unconscionable deprivations occurring within the

---

⋆ It was noted in Ch. 7 that even a desert-based sentencing system may invoke prevention for certain purposes, for example, in choosing between two sanctions of equivalent severity. Where, for example, a period of supervision is selected instead of a day-fine for an offender deemed to be in need of supervision and treatment, the requirements concerning acceptable penal content should apply also to that supervision.

walls. But these are unconscionable precisely because they lie so clearly outside the sanction's acceptable penal content.

Can this idea be carried over to non-custodial sanctions? We think it can. The first step would be to try to identify the acceptable penal content for such penalties. Certain kinds of impositions, we think, can be undergone with a modicum of self-possession, and thus would qualify. These would include deprivations of property (if not impoverishing); compulsory labour under humane conditions (community service, but not chain-gang work); and limitations of freedom of movement within the community. None of these deprivations involve the three evils cited above—destruction of the personality, demeaning ritual, or compulsory self-accusation. Clearly excluded would be measures which entail one of these evils, for example, self-accusation. This description is only suggestive, and we shall not try to offer a complete description of which kinds of non-custodial penalties do or do not meet the suggested standard. What we are proposing is a mode of analysis.

The analysis should be applied not only to the expressly punitive but also the supposed rehabilitative features of a programme. Deprivations administered for treatment are still penal deprivations, and can be no less degrading than deprivations imposed for expressly punitive ends. We would, for example, consider suspect a drug treatment programme in the community that involves compulsory attitudinizing. One may wish to persuade the offender of the evils of drug abuse and, for that purpose, deny him or her access to drugs or other stimulants, but if the offender is *compelled* as part of the programme to endorse attitudes about drug use that he or she does not necessarily share, that is bypassing his or her status as a moral agent.

Once the acceptable penal content of the sanction has been specified, certain ancillary deprivations may be necessary to carry the sanction out. Imprisonment, for example, involves maintaining congregate institutions and preventing escapes or attacks on inmates or staff. Segregation of some violent or easily-victimized offenders for limited periods may be necessary for such purposes, even if not appropriate as part of the the intended penal content in the first place.

Non-custodial sanctions also may require ancillary enforcement measures. Consider home visits. Such visits are not part of acceptable penal content: it is not plausible to assert that the penalty for a given type of offence should be that agents of the State periodically snoop into the offender's home. The visits can be justified only as a mechanism to help enforce another sanction that does meet our suggested standard of acceptable penal content. One such sanction might be community service, which we have suggested does meet the primary standard. To assure attendance at work sites by checking on excuses for absences, occasional home visits may be necessary—and, indeed, are part of the enforcement routine of the Vera Institute of Justice's pioneering community service project.[5]

Because home visits are thus justified only as an ancillary enforcement mechanism, their scope should be limited accordingly, that is, be no more intrusive than necessary to enforce the primary sanction. If home visits are ancillary to community service, they should occur only when the participant has failed to appear for work, and their use should be restricted to ascertaining the offender's whereabouts and checking on any claimed excuse for being absent. The less connected the visits are with such enforcement, and the more they are used to inquire into ulterior matters, the more they are suspect. Periodic general searches of an offender's home could not be sustained on this theory.

Are there any principled limits on ancillary enforcement sanctions, other than their being essential to enforce the primary penalty? Enforcement mechanisms that are grossly humiliating should be ruled out, even if needed as the enforcement tool for a particular kind of primary sanction. If X is a sanction with acceptable penal content in itself, but requires Y—a manifestly degrading measure—to enforce it, then X should be abandoned in favour of some other primary sanction that can be enforced less intrusively.*

What is meant, then, by grossly humiliating measures? It refers to measures that manifestly infringe our primary standard, i.e. that cannot be undergone by someone of reasonable fortitude with a modicum of self-possession. Think of the line between acceptable and unacceptable measures as a blurred one. The idea of 'acceptable penal content' requires that primary sanctions fall obviously to the permissible side of that line: that, for example, they not involve degrading rituals or compulsory attitudinizing. The ancillary measures may come closer to the line, but may not plainly fall foul of it. These are admittedly imprecise standards to deal with matters that resist precise definition. But we hope they offer some safeguard against the unacceptable types of penal intrusion.

4. THE RIGHTS OF THIRD PARTIES

The prison segregates the offender. The segregation, whatever its other ills, means that the rights of third parties are not directly affected. If A goes to prison, this does not restrain B's rights of movement, privacy, and so forth. Granted, B still suffers if he or she is attached to or economically dependent

* A similar analysis could be applied to electronic monitoring. One would ask, first, whether the measure is essential to enforce a primary sanction that meets our suggested standard of acceptable penal content; second, whether the scope of the monitoring is no more than what would be essential for that purpose; and third, whether the measure, when thus limited, remains grossly humiliating in the sense we have described. The answer may depend, in part, in what type of monitoring is involved. Is it, for example, designed only to verify the offender's presence at a prescribed location? Or does it also involve observation of the defendant's activities at that location, and to what extent so?

on A. But B, nevertheless, is not restrained. Non-custodial penalties, however, reintroduce the punished offender into settings in which others live their own lives. As a result, the offender's punishment may directly impinge on those others.

This third-party question is distinct from the issue of the offender's dignity, just discussed. Consider home visits again. Even if—for the reasons already suggested—such visits are not objectionable as degrading to the defendant, they affect the other residents, diminishing *their* sense of privacy. They themselves have done nothing to warrant State intrusion. Granted, when a person lives with someone else, he or she will inevitably suffer indirectly from whatever adverse consequences legitimately befall the latter person as a result of his misbehaviour. Nevertheless, there are some precautions that can be taken to help reduce the impact of non-custodial punishments on third persons.

Often, it is not the primary sanction but its ancillary enforcement measures that intrudes into the lives of third parties (to cite the previous example, home visits used to enforce community service). In such cases, the enforcement mechanism should be limited to helping insure that the *offender* abides by the conditions of the primary sanction. It should not be used—and becomes an unacceptable invasion of privacy when so used—to investigate the general extent to which others present abide by the law. When the defendant's home is visited to check on his or her excuse for being absent at the work site, that should not be used an occasion to gather evidence of illegal drug use by others in the apartment.

The impact on third persons should also be one of the criteria used in selecting non-custodial penalties. There may exist more than one sanction, of approximately equal severity, to choose from. Where the choice is available, the penalty should be preferred that intrudes upon third parties least. Suppose, for example, that the choice lies between a short period of home detention (enforced by unannounced home visits) or a fairly stiff schedule of community service (enforced by home visits to check the offender's presence, but only when he or she fails to appear at the work site). Suppose these sanctions are of approximately equal severity. If we conclude that the occasional home visits used to enforce community service are less disturbing to other residents than the more frequent unannounced home visits used with home detention, that would be reason for preferring community service.

This chapter provides more questions than answers. What we have done, at best, is sketch a tentative framework for dealing with questions of degradingness and intrusiveness. The issue badly needs further inquiry. Non-custodial penalties should not be permitted to become the means of introducing degrading penal routines into the community.

10

The Politics of Proportionality

THIS book, so far, has been devoted to the substance of sanctioning theory. It is time to turn to a different set of questions: political ones. What general political viewpoint, if any, is presupposed by a proportionality-based sentencing conception? Do proportionate sanctions tend to be more severe? Does desert theory divert attention from social ills? What settings are propitious, or unpropitious, for sentencing reform? Let me try to address these questions, in turn.

1. THE POLITICAL PEDIGREE OF PROPORTIONALIST SENTENCING THEORY

In penological thinking in the 1950s and 1960s, desert was considered a conservative—indeed a reactionary—idea. An example of this outlook is the 'Model Sentencing Act', published in 1963 by the National Council on Crime and Delinquency, a well-known American penal-reform organization.[1] According to the proposed Act, proportionality concerns were expressly to be excluded from the determination of sentence, and considerations of rehabilitation and incapacitation were to be relied upon exclusively. As a seemingly self-evident proposition, the proposed Act states in its first section that 'Sentences should not be based on revenge and retribution'.[2] However, the Act permits the sentencing judge, at his discretion, to impose a prison sentence of up to five years on any person whom he or she deems is likely to reoffend; and authorizes terms of up to thirty years those deemed especially dangerous. The drafters evidently did not see any risk to liberty involved creating such powers.

The so-called 'desert model'—that is, the renewed interest in the idea of proportionality of sentence—emerged in the 1970s among liberals, in part as a reaction against this kind of thinking. Fair sentences, it began to be argued, should reflect the gravity of the criminal conduct. Several years' confinement is not the kind of sanction which any offender deemed a bad risk should receive, but only those offenders whose actual criminal conduct is quite grave.

Although there was extensive philosophical writing on the subject previously,[3] my own 1976 work *Doing Justice*[4] was the first systematic exposition of desert theory in the penological literature. It was written on behalf of the Committee for the Study of Incarceration, a group (primarily of academics) with strong liberal sympathies. The principle of proportionality was offered as a means for *restricting* the State's authority to punish, particularly, as a way of limiting the use of severe sanctions. Predictively-based punishments were opposed, not only because they fail to reflect the degree of reprehensibleness of the criminal conduct, but because they permit intervention without limit into the lives of supposedly dangerous offenders. Sentence severities were to be sharply scaled back.*

Recent penological writings on desert generally preserve this tenor.[5]** The principle of proportionality is defended on grounds of fairness. Substantial prison terms are to be restricted to those convicted of seriously reprehensible criminal conduct. A penalty scale's anchoring points are to be set so as, generally, to reduce overall penalty levels. Proportionate sentencing is not offered as a means of reducing crime; indeed, desert advocates generally have been sceptical about how much crime rates can be made to respond to changes in sentencing policy. Absent an ambitious crime-reduction agenda, those writing in this vein have had little incentive to recommend increased reliance on prison sanctions.

The late 1970s and the 1980s, however, also witnessed the advent of a ferocious conservatism in criminal justice, especially in the USA. Politicians increased their emphasis on 'law and order' themes. Prison populations rose sharply. Conservative penologists such as James Q. Wilson and Ernest van den Haag urged 'realistic' sentencing strategies aimed at bringing crime rates better under control.[6] Because these developments were occurring at the same time as a renewal of interest in proportionality of sentence, some critics have suggested that proportionalism is part of a repressive strategy for dealing with crime. Desert theory has been denounced as a 'fundamentalist revival',[7] and its advocates described as 'architects of conservative theory'[8] or as proponents of the economic and social *status quo*.[9]***

---

* The work proposed that sentences for the worst offences should not exceed five years; see discussion of that proposal in Ch. 5, above.

** The writings to which I am referring are those of Andrew Ashworth, A. E. Bottoms, R. A. Duff, Nils Jareborg, Richard Singer, Martin Wasik, and myself. These writings, insofar as they address the rationale for proportionality, tend to rely on a censure-based rationale. Such a rationale permits the conventions through which the censure is expressed to be altered—so as significantly to reduce penalty levels.

At the outset of Ch. 2, I spoke of and argued against another conception of desert—one focusing on the 'unfair advantage' the offender obtains by offending. Such a conception (insofar as it offers any intelligible guidance on quanta of punishments) might point to a harm-for-harm criterion that could lead to quite severe sanctions for serious crimes. However, the unfair-advantage view has not had much influence upon sentencing theory.

*** For an extreme statement, see David Dolinko's recent 'Three Mistakes of Retribution' (1992).

The actual conservatives among penal theorists, however, have shown scant enthusiasm for the idea of proportionality. James Q. Wilson asserts that desert should be considered only in setting the broadest outer bounds on permissible punishments; and that sentence levels should be decided chiefly by deterrent and incapacitative concerns.[10] Ernest van den Haag once flirted with the idea of desert, but since has firmly rejected it.[11] (Indeed, he goes so far as to assert that convicted offenders have no legitimate equity-claims once they have submitted themselves voluntarily to the risk of being punished, through their decision to offend![12]) The reason for those theorists' scepticism is apparent enough (and indeed, it is the mirror image of liberals' reasons for support): the principle of proportionality limits the scope of crime-prevention strategies. Singling out high-risk offenders for extended confinement—a favourite theme of Wilson's[13]—would be ruled out as violative of ordinal proportionality. Exemplary deterrence strategies aimed at reducing the demand for drugs by penalizing users and minor sellers harshly—a favourite theme of the US drug warriors[14]—would also fall foul of proportionality requirements. If the primary aim of the sentence is seen as that of trying to stop crime, proportionality constraints are a mere hindrance.

The enactment of proportionality-oriented sentencing laws also has not been confined to conservative jurisdictions; indeed, it has been some of the less conservative places that have taken the lead. In the USA, it has been the sentencing commissions of two states—Minnesota and Oregon—that have made the most significant attempts to reflect proportionality concerns.[15] Both jurisdictions have had (by US standards) significant liberal traditions. And both states' guidelines were drafted so as to limit increases in sentence levels. When the much more law-and-order-minded US Sentencing Commission published its (rather draconian) guidelines for federal crimes, it expressly abjured a desert orientation.[16] Indeed, the commission's most influential member asserted that extensive reliance on desert rationale would have been inconsistent with the statutory mandate that multiple penal aims be relied upon, and would give insufficient scope to deterrence and incapacitation of criminals.[17]

In Europe, statutory sentencing principles rather than numerical guidelines have become the vehicle for instituting proportionality requirements. The lead was taken by two Scandinavian jurisdictions. In 1976—as interest in sentencing reform was only just beginning to develop in the USA— Finland amended its Penal Code to adopt a policy of proportionate sanctions.[18] The sentence, according to that statute, should be 'in just proportion' to the harmfulness of the conduct and the culpability of the actor in committing it. Predictive and rehabilitative concerns are largely excluded from the determination of sentence.[19] The Finnish enactment generated interest in neighbouring Sweden (once seen as the Mecca of rehabilitationism), and the Swedish government appointed a legislative commission to propose a new sentencing law.[20]★ The commission submitted its report in 1986, putting forward a more elaborate version of the Finnish law; that proposal, with modest changes, was approved by Parliament two years later.[21] According to the Swedish statute, the sentence should be determined primarily by the 'penal value' (i.e. seriousness) of the criminal conduct. The previous criminal record would also have a modest influence, but reliance on rehabilitative, incapacitative, or deterrent considerations is sharply circumscribed. The Finnish and Swedish enactments are of interest, because they involve countries which had little poverty, ambitious social programmes, and fairly humane correctional systems. That fact should give pause to those who assert that desert-oriented penal policies can gain influence only in rightist political environments.

2. DO PROPORTIONATE SANCTIONS TEND TO BE MORE SEVERE?

Another theme of critics has been that a policy of proportionate sanctions (whatever the aims of its original proponents) leads, in practice, to increased penalties. Barbara Hudson emphasizes this theme, citing evidence about higher penalty levels in California since its 'justice model' was first adopted in the mid-1970s.[22] The arguments, unfortunately, fall prey to the *post hoc ergo propter hoc* fallacy: since A has been followed by B, A must be the cause of B. No attempt is made to consider whether the increases occurred because of the influence of proportionalist sentencing conceptions, or due to other influences.

The use of imprisonment in California has doubtlessly increased substantially, since that state's determinate-sentencing statute took effect in 1977.[23] Citing such increases proves little, however, unless other critical questions are examined. Did the increases reflect pre-existing trends? Were they associated with features of the law which are proportionalist, or with ulterior features? Have increases of comparable magnitudes occurred in jurisdictions using a different technique of implementing proportionality? Have jurisdictions which eschewed the justice model succeeded better in avoiding penalty increases? When such questions are addressed, the supposed desert-severity link begins to fade.

California's prison populations were rising for seven years preceding the implementation of the determinate-sentencing law, and post-enactment increases continued at the same rate.[24] The increases resulted primarily from a sharp rise in prison commitments. A desert-based model should supply

---

★ At the time, the Swedish Penal Code provided that the sentence should, 'with an eye to what is required to maintain general obedience to law . . . foster the . . . offender's rehabilitation in society'.

standards for the decision whether to commit to prison—namely, ones that would limit incarceration to more serious offences.* The California determinate-sentence law never had such standards. The law only specifies the duration of confinement *if* the judge commits the defendant to prison, but (from the law's original enactment) left to judicial discretion the decision to commit.[25] Californian judges were thus free to send offenders to prison—and were, if anything, more ready to do so once they knew those individuals would no longer serve indeterminate, possibly lengthy terms. Not surprisingly, there has been a particularly significant increase in commitments for lesser felonies, such as burglary. So one wonders what California's increase in commitment-rates shows: is it that proportionalism leads to severity, or rather that *failure* to adopt a desert-based standard for the commitment decision leads (in a period of increasingly punitive attitudes) to more people going to prison?

Were desert responsible for increased severity, one would expect those states which rejected that idea to be less draconian. But have they been? The case of New York comes to mind. New York is comparable to California in demographics (both states have large cities with substantial low-income populations), and in the politics of law and order (crime is a big political issue in both states). New York, however, has never adopted any desert-based sentence-reform scheme, and has instead made extensive use of mandatory minimum sentences having a marked incapacitative flavour.[26] Contrary to Hudson's thesis, the absence of a justice model has in no way led to less punitiveness in New York: the state has experienced well over a trebling of its prison population in recent decades.

In 1985, New York considered moving toward more proportionate sanctions. A sentencing commission was established, and it proposed guidelines that would (albeit in somewhat crude fashion) have given the seriousness of the offence increased weight in deciding sentence. The proposal, although it called for moderate penalty increases, was roundly denounced as too lenient, and rejected. Leading the calls for rejection was the Manhattan district attorney, a former member of the commission. Prosecutors, he asserted, already had extensive powers to imprison under existing mandatory minima. The guidelines could only dilute their powers, by eliminating the ability to imprison routinely for second felonies, and by limiting discretion to go above the guideline ranges. If proportionalism is likely to lead to greater harshness, the point was apparently lost on the real get-tough advocates in New York.[27]

California's standards were statutory, and thus vulnerable to law-and-order pressures within the legislature. Dissatisfaction with statutory standards led, beginning in the late 1970s, to the development of interest in sentencing-commission guidelines. A commission, it was believed, would

* See Ch. 5.

be less exposed to such pressures, and would have more time and expertise to devote to writing the guidelines.

What has been the experience with sentencing guidelines? Some, most notably the US federal guidelines, are unashamedly draconian. The federal guidelines call for a halving of the use of probation and for a vast increase in prison commitments.[28] However, those guidelines—as just mentioned—expressly abjure a desert rationale.

What of guideline schemes developed with proportionality concerns expressly in mind? A notable instance is Minnesota's scheme, which has been in operation since 1980.[29] Minnesota's guidelines, in their original form, graded sentences according to the gravity of offences; and generally limited prison terms to crimes of actual or threatened violence and other serious offences, except in cases where the defendant has a fairly lengthy criminal record. During the first decade of their operation, the guidelines had considerable success in restraining the escalation of sanctions. The guidelines were designed to keep prison populations within prisons' capacity—and, during this period, largely succeeded in doing so.[30]

Even Minnesota is not exempt from panics over criminality, and one occurred in 1988 following a series of grisly sexual homicides in Minneapolis. Faced with heavy pressures from the legislature, the state's sentencing commission significantly increased the prescribed prison terms for violent crimes. However, the commission was able to exact its *quid pro quo*: the guidelines' criminal-history score was recalculated, so as to make it substantially more difficult to incarcerate property offenders with long records.[31]* The change was made to help maintain a degree of proportionality in the whole system, and to limit the increase in prison populations that otherwise would have occurred. Had there been no guidelines, I suspect that prison terms for violent crimes would still have been raised comparably or more; but there would have been no compensating reduction for property offenders. (Indeed, without the guidelines, there would have been no vehicle to help ensure that judges would make more sparing use of the prison for recidivist property offenders.)

In assessing the link between sentencing rationale and penal severity, a certain realism is needed. A sentencing theory cannot, Canute-like,** stop the waters from rising. Where the law-and-order pressures in a particular jurisdiction are sufficiently strong, punishments will increase, and no penal theory can prevent that. Increases in sanction levels are best understood with reference to the underlying political dynamics, and those have been apparent enough in much of the USA in the 1980s and early 1990s: increased public resentment over high crime rates, identification of crime in the public

---

* For fuller discussion of recent changes in Minnesota's guidelines, see A. von Hirsch and J. Greene, 'When Should Refomers Support Creation of Sentencing Guidelines?' (1993).
** With my apologies to King Canute!

mind with minority and lower-class groups, and an environment of political opportunism that has fostered official posturing about 'law 'n' order'. All a penal theory can do is offer reasons for limiting penalty increases. The question—to which I turn next—is whether desert, or utilitarian conceptions, can best supply such reasons.

3. GROUNDS FOR LIMITING SEVERITY: DESERT V. PENAL UTILITARIANISM

Does a proportionalist sentencing theory require more or less punishment? A major line of criticism has been that the theory provides no definite answer. All the principle of proportionality calls for is penalties graded to reflect the comparative gravity of offences, plus certain outer constraints on inflating the penalty scale. While the theory thus might *permit* a considerable reduction in punishment levels, it does not require this result; and indeed, may even allow a significant escalation of penalties. Proportionalism, the critics thus assert, comports with tough sentencing policies despite all its pretences to liberality.

I think I have answered this objection in Chapter 5. A desert theorist does not have to be agnostic on issues of anchoring the penalty scale. My suggested view would reject reliance on an 'optimizing' conception of prevention as the basis for setting the anchoring points; and offer an alternative conception as the basis for a decremental strategy that would reduce penalty levels substantially. I do not claim these conclusions to be distillable directly from proportionality requirements, but have suggested why they are consistent with the ideas of moral agency that also underlie the principle of proportionality.

The critic might reply that not all adherents to a policy of proportionate sentences would necessarily accept my view of anchoring the scale; and if they do not, the indeterminacy of desert theory remains. The objection, however, is now much weakened. The critic can no longer assert the impossiblity of developing a principled conception of scale-anchoring that points to reduced sanctions. It must be, rather, that some desert theorists might choose to reject such a conception, and thus continue possibly to support proportionate but tough penalties. However, this risk—of someone's choosing some aspects or versions of a theory but not others—would seem endemic to almost any conception of punishment one could devise.

Perhaps we should turn our attention from the details of the theory to its broader implications. What kind of sanctioning policy does proportionalism encourage? How would it compare as a way of constraining punitiveness with its major alternative of today, penal utilitarianism?

The most visible version of penal utilitarianism was once the treatment model: criminal-justice policy (including sentencing) was supposed to foster the rehabilitation of the offender. Faith in rehabilitation—at least, as the primary sentencing aim—has declined in recent decades.[32] With the decline of the rehabilitative ethic, penal utilitarianism has not disappeared, but merely shifted emphasis to other preventive strategies, notably, deterrence and incapacitation. If punishment cannot cure criminals, the thinking has run, more realistic ways of reducing crime exist: namely, intimidating potential offenders and restraining actual ones.

These deterrent and incapacitative strategies entail a particular risk of sanction escalation. There exists not only a danger of overall penalty increases (if a moderate dose of punishment fails to deter or incapacitate, why not try a bigger dose?); but especially, a danger of disproportionate escalation in sanctions for offenders targeted for special attention. What makes such strategies so worrisome is their common-sense appeal. While it is hard to imagine widely effective treatments, why cannot dangerous felons be intimidated or separated from the community? Doubts have by now been cast on how much currently-available preventive strategies can achieve such objectives.[33] However, the utilitarian can argue, what would prevent us from doing better if we continue to hone our crime-prevention skills?

Desert theory provides principled grounds for resisting such suggestions. The theory's guiding conception, of proportionality, is grounded in notions of equity. Preventive efficacy—the impact of sentencing reform on crime rates—is not the theory's primary criterion for success. If a desert model is implemented and does not lead to reduction in the incidence of crime, this is not a mark of the scheme's failure. Indeed, most desert advocates are (as noted earlier) rather pessimistic about achieving significant crime reductions through changes in sentencing criteria and practice. Sentencing reforms are to be evaluated, instead, in terms of their success in scaling the penal response to crime-gravity. Because the incidence of crime is not the primary criterion, the theory provides less temptation to escalate punishments in the hope of achieving preventive impact. Indeed, the Minnesota sentencing commission opted for desert over incapacitation in writing its standards, in part because it did not wish to present its scheme to the public as a crime-control device—and then, if crime rates continued to rise, face pressures to resort to tougher preventive medicine.[34]

The criteria for desert also rule out the scarier forms of penal utilitarianism, such as selective incapacitation. The latter strategy calls for imposition of substantially extended sentences on 'high risk' felons—namely those deemed likely to commit serious crimes repeatedly. It would utilize indicia of risk which have little bearing on crime gravity—such as prior arrests, drug abuse, lack of schooling, and unemployment.[35] A proportionalist conception of sentencing would preclude such an approach, because it relies on factors so ulterior to the blameworthiness of the criminal conduct, and also because (to achieve any significant preventive impact) it calls for such large disparity in the severity of sentences of persons deemed dangerous, as compared with others convicted of similar crimes.[36]

Could one argue against such utilitarian strategies, without relying on desert? One could try, but the arguments are weaker. Take selective incapacitation, again: Elliott Currie devotes much of his 1985 book to criticizing this strategy but is unwilling to invoke desert for the purpose.[37] Instead, he offers a number of technical arguments about why selective incapacitation is unlikely to work. (The answer to those arguments might be: provided the strategy is not going to cost too much, why not try anyway and see if it *does* work?) Missing from his discussion is the critical moral dimension: that there is something *wrong*, not just possibly ineffectual or counter-productive, in singling out some armed robbers for lengthy confinement while other robbers, convicted of comparable criminal acts, get much shorter terms.

One substitute response sometimes advocated recently is the reaffirmation of rehabilitation. Treatment, it is said, is a more humane notion than deterrence or incapacitation; and it is now widely understood that the prison is seldom a cure for offenders' criminal tendencies. A return to rehabilitation, supposedly, would pave the way for less harsh penal policies.[38]★ Ideologies cannot be revived at will, however. The rehabilitative ethic seemed so attractive in past decades, because people really believed that offenders could be cured routinely. That assertion scarcely rings true today: despite claimed successes with particular programmes, routine success with the bulk of offenders remains an elusive goal.[39]

The supposed intrinsic humaneness of treatment may also be doubted. The rehabilitative ethic offers few, if any, limits on how severe penal therapies may be. Rehabilitationism has had worrisome embodiments in the past, such as long, indeterminate terms of confinement for the sake of treatment. New rehabilitative techniques may also be less gentle than their proponents suggest. Drug treatments, especially, can be quite burdensome: compulsory residence in community-based facilities; elaborate treatment routines; and repeated searches and urine tests to detect continued abuse. Significant durations may also be involved: a recent summary of research claims that treatments need to run for ninety days or more, to show any effect.[40] The convicted drug user may, on a rehabilitative sentencing ethic, face such onerous interventions as these—instead of the more modest penalties a desert scheme should provide for such lesser crimes.[41]

A good test of the comparative merits of desert and utilitarian models concerns their applicability to intermediate sanctions. There have, as noted in Chapter 7, been two major problems with developing such measures. First, the sanctions are often employed on lesser offenders, because such

---

★ Cullen and Gilbert make this claim, as does Barbara Hudson. However, not all writers of comparable political orientation would agree: Thomas Mathiesen, for example, remains strongly critical of the treatment model.

persons are perceived as more co-operative recruits into the new programmes. Second, imprisonment may be invoked as the penalty for those who commit technical breaches of the terms of their community penalty, thereby ultimately increasing reliance on imprisonment. Desert theory does offer a principled response, as we saw: the more onerous non-custodial sanctions should not be employed for low-ranking crimes, and there should be substantial limits on breach penalties. Utilitarian theories, however, offer no such limits: breach, for example, could be treated as an indicator of increased risk that would justify incarceration.

4. ARGUMENTS ABOUT 'UNDERLYING ILLS'

Another cluster of criticisms of desert theory concerns its supposed failure to address certain fundamental ills, such as the underlying causes of crime or possible discrimination at earlier stages of the criminal process. Let me examine these objections.

1. Proportionalist sentencing doctrine sometimes is said to involve 'blaming the victim'.[42] Bad social conditions, assertedly, are actually responsible for crime (or at least a high incidence of crime). The desert rationale diverts attention from these conditions, by focusing on the supposedly wrongful behaviour of convicted offenders—whose crimes may, in large part, stem from their own economic victimization.

These are valid criticisms of the more conservative utilitarian theories, such as those of James Q. Wilson. In a much-quoted passage, Wilson argues that Government is not in a position to remedy fundamental social ills; and moreover, that it need not try to do so, because crime can more efficiently be dealt with through rigorous crime-prevention measures.[43] Thus his prescriptions for preserving public order do come at the expense of efforts to alleviate social injustice.

This is *not* a position that desert theorists need hold, however. A modern State should be capable of alleviating poverty and social disorganization. The European States which have made real (albeit admittedly costly) efforts in this direction have had some measure of success; and the USA has not so much been unable to reduce want as unwilling to undergo the necessary effort and expense. Nor would I concur with Wilson's view that it is possible, without trying to alleviate social misery, to reduce crime merely through this or that crime-prevention technique. It is no accident that Stockholm or Frankfurt are safer than Boston or Philadelphia, and the reasons why relate more to living conditions in those places than to criminal-justice efficiency.

Sentencing policy is not a good tool either for reducing criminality or promoting wider social justice. (Rehabilitative programmes, notwithstanding their proponents' rhetoric, also cannot rectify basic social injustices.) If we want a more equitable society, we will have to establish and pay for the requisite programmes of social assistance. That may also help shrink criminogenic conditions in the community, although one cannot be certain how much or when.

The sentencing of convicted persons, however, cannot wait until underlying social ills are remedied, nor can it be abandoned once they are addressed. Crime (including crime of a serious nature) will occur in any event. So the question is unavoidable: when convicted offenders face sentence (as more or fewer inevitably will), what guiding rationale will help assure that these decisions are made with the minimum of injustice? Addressing fundamental social ills (desirable and, indeed, essential as this is) cannot constitute a substitute for trying to make sentencing policy more coherent or fair.

The Swedish and Finnish experience illustrates this point. Both countries have for decades had a wide network of State-financed welfare supports, ranging from unemployment insurance through health care to child support. These programmes have doubtless helped eradicate want; and the wide sharing of prosperity may have contributed to the relative absence of violence and disorder. Crime has not disappeared in these countries, however, nor with it the need for an intelligible sentencing policy. Finland and Sweden switched to a proportionalist sentencing framework, because that was felt to provide clearer and more fair guidance. The adoption of these laws, however, was not premised on rejection of larger social welfare supports. If one were to ask the drafters of the Finnish or Swedish statutes whether these laws 'addressed' the country's remaining social ills, their answer would be certainly not—because that is the proper task of social-welfare measures, not criminal-justice legislation.

2. Desert theorists assert that—even if their scheme cannot cure social disadvantage—it at least does not leave disadvantaged offenders *worse* off. The factors determining crime-seriousness (and hence the severity of the penalty) concern primarily the conduct's harm and the actor's degree of criminal intent; social factors (such as employment status, education, age etc.) generally carry little weight. Desert is preferable in this respect to utilitarian strategies, for the latter allow consideration of social factors in a manner that put badly-off defendants in a still worse position when they face sentence: unemployment, lack of skills, unstable residence, and so forth can increase the penalty as indicia of risk.[44]

Barbara Hudson replies that such claims are misleading. Indigent defendants are at a disadvantage at all the stages of the criminal process preceding the sentence. These disadvantages—ranging from inadequate legal representation to a greater likelihood of pre-trial confinement—make them likely to be convicted on more serious charges than their more privileged fellows who engaged in comparable behaviour. Those more serious charges would, under an offence-based scheme, mean severer penalties. The supposed neutrality of desert is thus a sham. To offset their burdens at earlier stages, indigent offenders would have to have their social status expressly considered as grounds for extenuation. Hudson thus urges less reliance on desert, and more on a variety of 'non-legal' factors.[45]

One wonders, however, about the consistency of Hudson's viewpoint. If her critical account of the dynamics of criminal justice is correct, then what makes her think that, were desert rejected, the 'non-legal' factors introduced would be ones to her liking? Would not the preferred non-legal factors be precisely those targeted against the so called 'dangerous classes': namely, incapacitative concerns that treat social deprivation as a sign of increased risk?

A consistent perspective is essential here. One might begin with something approaching an ideal perspective: what sentencing policy should look like in a reasonably just (or at least, not manifestly unjust) society. In such an environment, much may be said for desert theorists' view that punishment ought to reflect the blameworthiness of the conduct, and not ulterior social-status factors. There might be some exceptions, for those social factors that seem to bear directly on the blameworthiness of the criminal conduct. Martin Gardner[46] has suggested that extreme want reduces the culpability of the actor in committing a criminal act, and thus should be treated as an extenuating circumstance. Assuming the society had only small pockets of deprivation, such exceptions would need to be invoked rather rarely. Next, one could focus on the grimmer realities of the real world of which Hudson speaks, in which deprivation is more extensive and the deprived are at a real disadvantage in the legal process. There, however, it may be hazardous to urge heavy reliance on 'non-legal' factors. If crime rates are significant, there will be great difficulty obtaining approval for poverty-based mitigating factors, that diminish punishments for large numbers of low-income convicted offenders. Increased use of social-status factors is likely to produce just the opposite result: the entry of risk-considerations that make poor defendants even worse off. What one cannot do, however, is try to have it both ways: criticize the use of desert factors from a 'realistic' perspective, and at the same time urge adoption of other, 'non-legal' ones for immediate use, from a utopian viewpoint.

5. THE SETTINGS FOR SENTENCING REFORM

Sweeping factual generalizations have become suspect these days. Nowhere is such scepticism more warranted than in the politics of criminal-justice reform. Severity trends following enactment of proportionalist sentencing law or regulation have varied greatly from place to place. In Finland, the enactment of its 1976 sentencing-reform statute was followed by significant reductions in prison populations over the ensuing decade and a half.[47] In

Minnesota, the story was more complicated, as we have seen. In California, a flawed determinate-sentencing statute had little success in restraining the rise of prison populations, in a political atmosphere in which crime issues had become increasingly visible and contentious.

The prospects for success of recently-enacted reforms, and of future ones, will also vary with the institutional and political structures of the jurisdiction, and with the manner in which the reform is carried out. Proportionalist sentencing schemes do not have an intrinsic, inevitable dynamic, as critics seem to assume. To see how much the setting and the details count, it might be useful to contrast two recent, proportionality-oriented reforms: the 1988 Swedish sentencing reform law, and the English Criminal Justice Act of 1991. The Swedish law was steered through Sweden's Parliament by the then-ruling Social Democratic Government. In England, legislation with a comparable philosophy, and a somewhat analogous structure, was put forward by a Conservative Government. Both laws have entered into force too recently to permit an assessment of their impacts. Yet it surely is too simple to suppose that the sponsorship alone will be decisive—that Sweden's 'Social Democratic' law will lead to reduced punishments and England's 'Conservative' one to increased sanctions. What is likely to matter more are such factors as the following.

1. A jurisdiction's penalty levels, we saw, respond to the intensity of 'law and order' demands: the more political debate focuses on the supposed need for tough punishments, the more difficult it becomes to resist increases.

The Thatcher Government periodically made ferocious pronouncements about crime, but a general escalation in punitiveness was not one of its priorities. No ambitious proposals to raise penalty levels were put forward. Legislation to expand the rights of defendants in police custody was adopted in 1984.[48] The 1989 White Paper on sentencing expressly endorsed the idea of reduced reliance on imprisonment for property offenders.[49] The conservatism (indeed, extreme conservatism) of Thatcherite policies focused on other areas: welfare, local taxation, education, and union rights.* The passage of the 1991 Criminal Justice Act doubtless was facilitated by the fact that law-and-order was no big political issue.

However in the spring of 1993 (just a few months after the Act came into force and as this book was going to press), the mood has been changing. Law-and-order is becoming a theme for both major parties, with the Prime Minister proclaiming the need for a 'crusade' against crime. Demands are being heard from politicians and judges to alter the 1991 Act, and there are indications that some important provisions* may be watered down or eliminated.

---

* Unit fines may be eliminated, as may be Section 29 of the Act which imposes certain statutory restrictions on consideration of prior convictions.

In Sweden the situation has been somewhat different. In the 1991 national elections, the conservative coalition that ultimately won had to be cautious in criticizing the country's broad network of social guarantees, as these had extensive popular support. Law-and-order was one of the few issues which opposition politicians could use to challenge the ruling Social Democrats, and use it they did. The new Government has favored raising levels of imprisonment—and has quite recently brought about a modest rise through a change in the parole rules, noted below. However, fiscal constraints impede adoption of large sentence increases,[50] and—despite grumblings from some conservative politicians about repeat property offenders—there has been no concerted effort to alter the basic structure of the 1988 sentencing law.

2. The small print (that is, particulars of the legislation) count, as do judicial traditions. The Swedish statute sets forth its proportionality requirements at some length—and the statutory requirements are supplemented by an extensive legislative history that judges are expected to consider. The English statute is more loosely drafted: for example, the provision requiring imprisonable crimes to be 'serious' is set forth in quite general terms.[51] It is accompanied by another provision, intended to reduce reliance on the previous criminal record, which is also drafted somewhat obscurely.[52] The impact of those provisions on rates of imprisonment will depend on how seriously they are taken by the courts. A more detailed and clearer directive might have had more certain effects.

3. The provisions ancillary to sentencing may have a considerable impact—particularly, those regarding parole. The English law postpones parole eligibility from one-third to one-half of the sentence—but then requires automatic release after one-half for shorter-term prisoners.[53] The effect of these changes remains to be assessed. Sweden has also considered a postponement of parole eligibility—from one-half to two-thirds of the sentence. The legislative committee proposing the 1988 statute recommended, in order to offset the effects of this change, a scaling-down of statutory minima and maxima.[54] The new Swedish Government has opposed this recommendation, and has just obtained enactment of the postponement in the parole release date, without any reduction of the statutory minima. The effect of this change in the parole rules will be, for the sentences affected, an approximately one-sixth increase in durations of confinement. (However, the change relates only to prison terms between two months and two years—so that the majority of prison sentences, which in Sweden are below two months, will not be affected.)

In assessing the impact of the Swedish and English laws, developments throughout Western Europe are also worth bearing in mind: particularly, trends toward increasing punitiveness. The recent opening to the East has produced its dislocations—and with these, the possibility of increased crime rates, and the probability of increased public fears of crime. Even countries such as Germany and Austria—which to date have been rather successful in reducing their levels of imprisonment[55]—may begin to find themselves confronted with demands for changed policies. To the extent that European countries come to face sufficiently strong political pressures for tougher crime policies, no penal theory (desert included) can block such trends.

What a punishment theory can do is avoid providing an easy justification for penalty increases; and to help make their impact less inequitable. On these counts, a proportionalist conception of sentencing strikes me as preferable. It is less concerned with the supposed crime-prevention effect of this or that penalty, and more with fairness in penalties' distribution. This makes it harder to defend singling out particular types of offences or offenders that evoke public concern at the moment—drug abusers, burglars, or whatever—for disproportionate increases in sentence. A modest pro rata penalty rise has the same impact on prison-population statistics that large penalty increases focused on a few offence or offender types. Yet in terms of the impact on the lives of those involved, the latter is much more devastating.

Issues such as these are eminently worth exploring more fully. My basic point, though, is that if we are to talk about the politics of sentencing reform, we need to take politics seriously and look at the political environment of the particular jurisdiction in question. Generalities will not help.

Epilogue

The Aspiration to Penal Justice

THE assumption of this book has been that a sentencing system should seek to be just—or at least, to be as little unjust as possible. Claims about fairness, not crime-control effectiveness, underlie the requirements of proportionality, as the reader has seen. To aim at fairness would seem to be an uncontroversial objective for a decent and humane sanctioning system. Yet this aim has elicited a variety of objections, the common theme of which is that achieving penal justice lies beyond the capabilities or legitimate aspirations of a modern State. These objections can be summarized as follows:

1. A legal system must operate with rules, and proportionalist sentencing schemes (with their emphasis on treating similar crimes similarly) particularly require rules. Legal rules, however, inevitably treat *dis*similar cases similarly, and thus create only the appearance of equity.

2. A desert rationale assumes that criminal conduct is wrongful, and hence deserving of censure expressed through punishment. It thus presupposes consensus about what is deemed right and wrong, and a criminal law capable of giving authoritative expression of that consensus. In a modern industrial (or post-industrial) society, however, these conditions do not exist.

3. Many criminal offenders come from deprived segments of the population. The blameworthiness of such persons' conduct is in doubt, given their limited opportunities to abide by the law. To speak of underpriviledged offenders as deserving of punishment, and to try to calibrate their penalties on the basis of how much they might deserve, is a fiction.

4. Proportionality, with its emphasis on equal treatment of those convicted of equally reprehensible conduct, conflicts with another value: that of parsimony. Where such a conflict occurs, a humane penal system should prefer parsimony.

Each of these claims, especially the last two, warrant more extended discussion than this brief epilogue can provide. Let me just state succinctly why I do not find these objections convincing.

1. TREATING DISSIMILAR CASES SIMILARLY

Ordinal proportionality requires that equally serious conduct be punished with approximately equal severity, and that unequally serious conduct be penalized so as to reflect the differences in gravity involved. To implement this requirement, sentencing norms or standards are needed—to ensure that individual sentencers make consistent judgements of the gravity of crimes and the commensurate severity of punishments. Such norms, however, necessarily will be aggregative: they will lump together cases which, on more careful inspection, might involve differing harm or culpability. How aggregative they are depends on the form of the rules. Numerical sentencing guidelines tend to be quite strongly so: they impose a uniform punishment within an offence category or subcategory, save in special cases of aggravation or mitigation. Statutory sentencing principles permit more differentiation, as they permit development of a case-law jurisprudence that distinguishes degrees of harmfulness or culpability within an offence category.★ But even such a jurisprudence will (if it is to avoid unmanageable complexity) have to treat some possibly distinguishable cases similarly: legal rules are capable only of a certain degree of refinement.

Some writers, such as Michael Tonry in a recent essay,[1] seize upon this fact to argue that no proportionalist sentencing scheme is capable of doing true justice. Because the standards for proportionality rely on legal categories, they will lump morally dissimilar cases together. By so doing, the standards will merely *purport* to be fair, but in fact be unfair.

This argument puzzles me. Meeting proportionality requirements is a matter of degree. A sentencing system can, to a greater or lesser extent, punish comparably blameworthy conduct similarly. No sentencing system can *completely* succeed in this respect, because of the aggregative features of sentencing rules. Nevertheless, a carefully-worked out system of standards can achieve proportionality to a reasonable degree—which is, I think, all that can sensibly be aimed at.

What seems to be behind Tonry's objection is an all-or-nothing approach. Either proportionalist sentencing schemes can achieve complete justice; or else, if that is not feasible (as it admittedly is not), then the pretence of being fair should be abandoned. Sometimes, this is stated by saying that preventionist sentencing schemes may not be particularly fair, but at least they are honest in their repressive aims and do not pretend to be just. But the all-or-nothing approach—in sentencing as well as in other areas of social policy—simply leads to ignoring concerns about justice or making them largely marginal or symbolic. When this occurs, policies may

---

★ See at the end of Ch. 8, above.

well emerge that are not merely less-than-perfectly-fair, but plainly unjust.

The concern about aggregating dissimilar cases has a legitimate role to play, but it is a different one: that of providing a critique of particular schemes for sentencing guidance. It doubtless is a drawback of numerical sentencing guidelines such as Minnesota's that they are highly aggregative, and give so little scope within broad offence categories to differentiate on harm or culpability grounds.[2] What that suggests is that standards should preferably be more sensitive (albeit they cannot perfectly be so) to such variations in gravity.

2. PUNISHMENT AND MORAL CONSENSUS

The next objection concerns the absence of moral consensus. Traditional societies, it is said, have a considerable degree of agreement about what is right or wrong. In matters of dispute, there are authoritative figures to consult. In such a society, there would be little difficulty grading penalties to fit the supposed degree of reprehensibleness of the conduct. Our own societies are not so constituted, however. There are contending groups having differing outlooks about how people should behave. The State also carries no particular authority to resolve ethical disputes. How then, the objection runs, can the degree of various acts' wrongfulness be determined, sufficiently to construct a desert-based sentencing scheme?

That dissensus exists in modern societies on some legal and ethical issues is undeniable: witness the ongoing debates about abortion and drug use. The criminal sanction performs poorly when it tries to regulate such areas of dissensus, which is one (but not the only) reason why it should enter such areas with caution. However, if we turn our attention to the core areas with which the criminal law should deal—the prohibition of victimizing acts of force, theft, and fraud; and the enforcement of certain basic duties of citizenship such as the payment of taxes and the preservation of the environment—there appears to be a greater degree of agreement.*

The objection also relies on a faulty model for how the comparative gravity of crimes is to be determined. It assumes that the State, when it tries to rank crimes in seriousness, must try to mirror (or else to give authoritative expression to) the wider ethical norms of the society in their full complexity. But perhaps the task is a more manageable one. Typical victimizing crimes, I suggest in Chapter 4, can be assessed in their seriousness by analysing the harm and culpability involved. Harm can be gauged by the typical impact of the conduct on a person's living-standard; culpability, by the conduct's degree of purposefulness or carelessness.

It is the State, not society as a whole, that punishes, and through its

* See Ch. 4, above.

punishment conveys penal censure. What the State needs, in deciding how much to punish, is reasons supporting its judgement that one kind of conduct is more serious than another. The judgements of harm and culpability of which I have just spoken furnish such reasons. It is not necessary to seek and try to reflect precisely a larger social consensus.

Matters admittedly become more complicated when one leaves the core area of victimizing offences, and goes to crimes (such as drug offences) the wrongfulness of which is in dispute.* Here, analysis is impeded by the lack of an adequate theory of criminalization—a theory of when, and under what circumstances, conduct may be deemed sufficiently reprehensible to warrant the blaming response of the criminal sanction. But if assessing the gravity of these crimes is more difficult, that would seem to me to be a strength, not a weakness, of a proportionalist sentencing theory. When it is to be doubted whether and why the prohibited conduct is wrong, it should come as no surprise that its degree of gravity is hard to gauge.

3. PROPORTIONALITY AND SOCIAL DEPRIVATION

No modern society has succeeded in eradicating social deprivation. In the United States, poverty and social disorganization exists on a massive scale, and affect a large segment of minority populations; in Britain, want is prevalent in many lower-class neighbourhoods; and even in European countries with extensive social insurance systems such as Germany and Sweden, pockets of misery (and in former East Germany, more than just pockets) remain. These persisting ills raise the issue of 'just deserts in an unjust society'. The distinctive feature of proportionalist sentencing schemes is that they purport, not merely to help prevent undesired conduct, but to punish fairly. How can punishment be fair in a society that is not itself equitable?**

I addressed this issue nearly two decades ago, in the last chapter of *Doing Justice*,[3] but made the question still harder than it had to be. There, I espoused the unfair-advantage theory of penal desert, described and criticized in Chapter 2 of the present volume. This view is particularly vulnerable to 'unjust society' objections. Unless it is in fact true that a country's social system succeeds in assuring the provision of mutual benefits to all citizens, the lawbreaker does not necessarily gain (or gain much) from others' law-abiding behaviour; and hence is not taking unfair advantage of others' self-restraint when he offends.

However, I have become convinced (as noted in that chapter) that the

---

* For the ethical issues involved in the drug prohibition, see D. Husak, *Drugs and Rights* (1992).

** See also the discussion in Ch. 10 above, of proportionalist sentencing theory's supposed inability to remedy underlying social ills.

unfair-advantage theory is unsound and fails to provide meaningful guidance on how much punishment is deserved. More persuasive, I suggest, is a censure-based rationale for proportionality, that focuses on punishment's role in conveying disapproval of criminal conduct. Such a censure view does not depend directly on how much the actor supposedly benefits from others' compliance with the law. Instead, it focuses on how much harm the conduct does or risks, and how culpable the actor is for the harm. The unjustly-treated actor may still be to blame for his misdeeds— particularly, when he injures others who are themselves not responsible for his ill-treatment. Moving to a reprobative rationale for punishment alters the social-deprivation issue. It is no longer enough to point to social injustices; one needs to explain why and to what extent the fact of social injustice might undercut claims about the harmfulness, or about the culpablity, of the conduct.

First, does the fact of social deprivation affect the conduct's harmfulness? The so-called 'left idealist' view of crime maintained that criminal conduct was not so much intrinsically injurious as it was a social construct serving certain class interests—particularly, dominant social classes' interest in keeping deprived populations under control. By recognizing the skewedness of existing social arrangements, the argument ran, the actual harmfulness of conduct denominated as criminal could be called into question.[4]

For some types of crime (e.g. prostitution, perhaps drug offences), this view has considerable appeal. However, for other, more predatory kinds of conduct, it seems roseate. This has now been recognized by the so-called 'left realists', who point to the realities of crimes of assault, burglary, and theft in poor neighbourhoods—to the injury done not only to individual victims but to the social bonds of the community.[5] Whatever social injustice might otherwise do, it does not diminish (indeed, it may increase) the harmfulness of common victimizing crimes.

The criminal law doubtlessly tends—through its prohibitions of disruptive conduct—to uphold existing social orders, whether good or bad. But, as Richard Sparks[6] and David Garland[7] have pointed out, it also has important practical effects in helping to protect ordinary people's lives, safety and possessions. It is far from clear that the law's repressive functions predominate everywhere over its protective ones.

Second, does the fact of social deprivation eliminate or reduce the culpability of offenders? Restricted economic and social opportunities could make it harder for a person to resist temptations to offend; and this increased difficulty of compliance, arguably, makes the person less culpable if he fails to comply. But how much greater must the difficulty of compliance be, before a claim of reduced culpability is warranted? And where warranted, how much is culpability reduced?

If such questions are pursued, they might point to a mitigating factor for those who have suffered serious social deprivation of a kind that would make compliance very much more difficult.★ What is not supportable, however, is the broad inference that critics wish to draw: that social deprivation undercuts the legitimacy of proportionate sanctions in general. Even if misery or want were to mitigate the culpability of its greatest sufferers, it is far from obvious why, or how, it renders all or the bulk of law-breakers beyond blame. The latter position would call for the unlikely assumption that deprivation is so extreme and persuasive that all offenders (or all underpriviledged offenders) have virtually no choice at all on whether to offend.

Nevertheless, this issue is a troublesome one. It would not be easy, even in theory, to determine when deprivation is sufficiently grave, and sufficiently related to the conduct at hand, to warrant mitigation. Further problems would be encountered drawing any such mitigating factor with sufficient clarity to make it workable in application. Getting such a factor adopted for a sentencing system is likely (as pointed out in the previous chapter) to encounter formidable political obstacles, in any jurisdiction other than one in which poverty is rare. Such difficulties should not surprise one. A sentencing scheme, on any theory, scarcely can compensate for the effects of wider social ills. A proportionalist sentencing theory at least permits this issue of social deprivation to be raised. The more proportionality is disregarded, the less the offender's culpability will count—and hence the less reason there would be for even suggesting that impoverished offenders might be entitled to less punishment.

Can anything be done, then, about the question of social deprivation? If, on one hand, deprived persons (or some of them) might arguably be entitled to something less than the full measure of the prescribed punishment, but if, on the other, granting such special mitigated treatment is likely to encounter serious theoretical or practical obstacles, one possible solution remains: to to keep such matters in mind, when setting a scale's anchoring points. This would support the recommendations I have made in Chapter 5, for a substantial reduction of overall punishment levels.

★ A few offenders may be in the direst need, and commit the offence to meet the need, i.e. cases akin to the oft-cited one of the starving person who steals the loaf of bread. Such persons, however, should be exonerated by the substantive criminal law's necessity defence (see Fletcher, *Rethinking Criminal Law*, 774–98, 818–29). More frequently the issue is something else: the person's restricted options may make compliance not virtually impossible but harder. In such situations, it is not arguable that the offender is *totally* without fault. The more plausible claim is Martin Gardner's (referred to in Ch. 10, above), that the added difficulty of compliance renders such persons *less* culpable. If, however, one asks why the person's culpability is reduced, a mere general reference to his deprived status would not suffice. One would need to cite particular reasons why that deprivation, in the circumstances, was of such a nature as would render compliance especially difficult. Undertaking such an analysis, however, will narrow the scope of mitigation: social deprivation could be indicative of reduced culpability when, and only when, certain specified conditions apply, relating to the situation of the offender and the character of the offence.

4. 'PARSIMONY OVER PROPORTIONALITY'

Does proportionality conflict with parsimony? Some critics assert it does. Michael Tonry, in a recent essay, asserts that 'proponents of strong proportionality conditions necessarily prefer equality over minimization of suffering'.[8] The preference, he asserts, should be reversed: parsimony should come first. If it does, selective reductions in punishment should be permissible even when they infringe ordinal proportionality constraints.

Why, and how, should proportionality give way to parsimony? Tonry offers two kinds of cases in support of his thesis. On closer scrutiny, however, these cases have narrower implications than he wishes to draw; and his thesis loses its credibility when applied more widely.

Tonry cites the case of a former drug abuser from a poor background who, against considerable odds, has worked his way up to be a mechanic in an automobile service station; and who (save for his current offence) seems well on his way to becoming a responsible citizen. He contends that such an offender should receive a more parsimonious sanction than the full measure of that which would be proportionate for his crime.[9]

There is, admittedly, something appealing about this case. Even the Swedish Committee on Imprisonment, which drafted Sweden's current, desert-oriented sentencing law, proposed a mitigated sentence for cases where the offender has improved his situation through his own efforts.[10]* If one asks why this kind of case is appealing, the reasons do not seem to be purely consequentialist. The Swedish Committee did not propose reducing the punishment for any offender who has become a better risk, but only for the offender who has brought about the improvement through his *own* efforts. The Committee's idea seems to have been that there is something specially meritorious in his having made such efforts to become more law-abiding.**

Should this be a mitigating ground? There might, as I mentioned in my discussion of 'hybrids' in Chapter 6, be downward departures from ordinal desert constraints in exceptional cases warranting a special degree of sympathy. What remains debatable is whether this particular kind of case should qualify.[11]*** Were one to conclude it should and grant the reduction in

---

* Committee's draft provides for a reduced sentence '. . . where through the actor's own efforts, a considerable improvement has occurred in his personal or social situation that bears on his criminality'.

** Part of the appeal of the mechanic's case may also reside in the issue just discussed, of the fact of his impoverished background.

*** To make a case for such an exception, two problems would need to be addressed. One is that of clarifying the normative grounds for the exception. Improving one's personal situation against odds may be meritorious, but why is it the kind of merit that should bear on how much one is to be punished? Some kinds of merit, I would assume, should not be relevant: my punishment for a crime should not diminish because I give to charity or have rescued a person

punishment, however, that would not warrant Tonry's conclusion that proportionality constraints should be watered down generally. In this connection, it is worth recalling that the Swedish Committee dealt with the matter by providing an explicit exception to its general rule of proportionate punishment.

In another essay,[12] Tonry cites a different situation where he thinks 'undeserved' downward departures from the tariff would be warranted: namely, when the tariff itself is unjustifiably inflated. The case he cites is that of the US federal sentencing guidelines. Those guidelines now prescribe incarceration for almost all crimes, other than those having the lowest offence rankings. To reduce imprisonment rates he proposes a first-offender exception: that the court should be authorized to impose a non-custodial penalty on any offender without previous convictions, except where the offence is violent or otherwise 'serious'. The enabling statute explicitly authorizes the US Sentencing Commission to create such an exception[13]—making the step easier for the Commission to defend politically.

In terms of proportionality, such a sweeping first-offender exception is not defensible. While the absence of a prior record may be mitigating to a limited extent, the current offence should bear primary weight in deciding the punishment.* Under Tonry's proposal, that would no longer be true: the criminal record would chiefly become determinative.

The Commission's tariff, however, is deeply flawed. It was not designed with proportionality in mind. By prescribing the severe sanction of incarceration for all but the least serious crimes, it almost surely infringes cardinal proportionality.** Seen in this light, Tonry's first-offender proposal for the federal guidelines is not a case of favouring parsimony over proportionality. The tariff fashioned by the US Sentencing Commission in no wise satisfies proportionality requirements. Creating a first-offender exception is a way of reducing some of the tariff's excesses. One might doubt whether Professor Tonry's suggested solution is the best. If any prospects exist of less inflamed law-and-order politics at the federal level, it would seem preferable to try to reduce the tariff as a whole, rather than create this somewhat arbitrary exception for first offenders. But that is a

in distress. Why is the mechanic's kind of merit different? The other problem concerns what German penal theorists call *Lebensführungschuld*: of looking at how an offender conducts his civil existence in deciding how much he deserves to be punished. The most obvious risk of introducing such matters will be to prefer those defendants whose lifestyles fit better with middle-class mores. For reasons of personal autonomy, the State should not use the criminal law to seek to regulate a person's work habits directly. If so, is it not problematic to do so indirectly through the factors relied upon in deciding sentence?

   * See discussion and references in Ch. 7.

   ** For discussion of the federal guidelines, see Ch. 10; for the requirements of cardinal proportionality, see Ch. 5.

matter of trying to alleviate the worst features of a defective system—not of the jursiprudence of proportionality versus parsimony.

For Tonry's thesis of the pre-eminence of parsimony to be correct, it must hold even in a penalty structure which does meet proportionality requirements. Let us imagine a scheme in which sanctions are graded according to offence gravity, and in which sanction levels are set well below those of the US Sentencing Commission's guidelines. If parsimony trumps proportionality, then it should be obvious, even with such a scale, that selective penalty reductions would be desirable. But who should be the beneficiaries? Would it be appropriate to give every fourth defender a lesser sentence than his fellows? What of something not quite so obviously arbitrary, such as a sharp penalty reduction for first offenders? Such a reduction would create a large disparity in the treatment of first offenders and recidivists. The disparity, as just noted, could not be justified on grounds of desert. It also could not be defended on grounds of risk, without a defence of the merits of such an incapacitative criterion.[14*] If parsimony is the purported justification, does this not amount to saying that the reduced penalties for the first offenders are a good thing merely because those persons will be punished less? That claim, standing alone, does not strike me as especially convincing—especially as the alternative exists by trying to proceed with the decremental strategy of making rateable rather than selective reductions.

Parsimony concerns the laudable goal of reducing penal suffering. It points to a lowering of penalty levels—something that proportionate sanctions can achieve, by moving the penalty scale's anchoring points downward. If selective penalty reductions are made instead, it needs to be explained why the particular individuals or groups involved are favoured. Parsimony, however, does not tell us who should benefit from penalty reductions, how much, and why so. Why should the two types of offender Tonry mentions—the mechanic and the first offender—be the chosen ones, instead of other persons? Is it merely that it is politically easier to reduce penalties for these persons—and if so, does this ground suffice? Or is it that it is somehow right or fair to select them? Once the fairness of the selection criterion is addressed, however, there is no avoiding the question of when, if ever, it is appropriate to deviate from proportionality to favour certain offenders. 'Parsimony over proportionality' has appeal as a slogan, but it is not illuminating when given closer scrutiny.

---

* A simple distinction between first offenders and recidivists would not even serve particularly well as a risk criterion, because the number of prior convictions (as contrasted to prior arrests and other social-history information) is not sufficiently correlated with recidivism.

Notes

CHAPTER 1

1. See A. von Hirsch, *Doing Justice* (1976).
2. For text and commentary to the Act, see M. Wasik and R. D. Taylor, *Blackstone's Guide to the Criminal Justice Act of 1991* (1991); see also A. Ashworth, *Sentencing and Criminal Justice* (1992).
3. See excerpts from Bentham and discussion in A. von Hirsch and A. Ashworth (eds.), *Principled Sentencing* (1992), ch. 2.
4. Ibid.
5. Ibid., ch. 1.
6. See e.g. N. Morris, *Madness and the Criminal Law* (1982), ch. 5.
7. For some of the literature of that period, see e.g. American Friends Service Committee, *Struggle for Justice* (1971); M. Frankel, *Criminal Sentences* (1972); D. Fogel, *We Are the Living Proof* (1975); A. von Hirsch, above n. 1; Twentieth Century Fund, Task Force on Criminal Sentencing, *Fair and Certain Punishment* (1976); D. M. Gottfredson, L. Wilkins, and P. Hoffman, *Guidelines for Parole and Sentencing* (1978); A. Blumstein, J. Cohen, S. Martin, and M. Tonry (eds.), *Research on Sentencing* (1983).
8. For an analysis of the guidelines and their impact, see A. von Hirsch, K. Knapp and M. Tonry, *The Sentencing Commission and its Guidelines* (1987), chs. 2, 5, and 8.
9. F. E. Zimring and G. Hawkins, *The Scale of Imprisonment* (1991).
10. For a critical analysis of the US Federal Sentencing Guidelines, see A. von Hirsch, 'Federal Sentencing Guidelines' (1989).
11. For discussion and citations, see von Hirsch and Ashworth, above n. 3, ch. 6.
12. The Minnesota and Oregon sentencing commissions have been discussing extending their sentencing guidelines to non-custodial penalties, albeit not yet with any definitive results. Delaware has standards in operation governing the use of such penalties, although their rationale is less easy to pinpoint. Guidelines on the use of unit (or 'day') fines have been developed in several counties, including those in which Phoenix, Arizona, and Des Moines, Iowa are located. The standards, which are based on a model developed by the Vera Institute of Justice for Staten Island, New York, rely primarily on the seriousness of the criminal conduct in fixing the number of fine units assessed. For a description of the Staten Island model, see J. Greene, 'Structuring Criminal Fines' (1988).
13. See e.g. R. A. Duff, *Trials and Punishments* (1986); N. Lacey, *State Punishment* (1987); A. Ashworth, 'Criminal Justice and Deserved Sentences' (1989); J. Braithwaite and P. Pettit, *Not Just Deserts* (1990); N. Walker, *Why Punish?* (1991); A. Ashworth, *Sentencing and Criminal Justice* (1992).
14. See e.g. M. Wasik and K. Pease (eds.), *Sentencing Reform: Guidance or Guidelines?* (1987).
15. UK Government White Paper, *Crime, Justice and Protecting the Public* (1990).
16. Von Hirsch, Knapp, and Tonry, above n. 8, ch. 5.

17. For discussion of where the 1991 Act differs from the White Paper, see A. Ashworth, above n. 13, chs. 4, 9, and 10.
18. Von Hirsch, above n. 1, ch. 8; A. von Hirsch, *Past or Future Crimes* (1985), ch. 3.
19. See e.g. Duff, above n. 13; J. Kleinig, 'Punishment and Moral Seriousness' (1992); I. Primoratz, 'Punishment as Language' (1989); U. Narayan, 'Adequate Responses and Preventive Benefits' (1993).
20. See Braithwaite and Pettit, above n. 13, 148–150; and Walker, above n. 13, 102–3.
21. P. Robinson, 'Hybrid Principles for the Distribution of Criminal Sanctions' (1987).
22. See e.g. N. Morris and M. Tonry, *Between Prison and Probation* (1990).
23. See Criminal Justice Act 1991, ss. 6(1) and 6(2)(b). See also A. Ashworth, above n. 13, ch. 10.
24. For citation of such literature as exists, see Ch. 7, below.
25. See e.g. B. Hudson, *Justice Through Punishment* (1987).
26. The first two of these three assumptions are discussed also in von Hirsch, above n. 1, ch. 1.
27. For fuller discussion, see A. von Hirsch, 'Equality, "Anisonomy", and Justice' (1984), 1105–7.

CHAPTER 2

1. See e.g. D. E. Scheid, 'Kant's Retributivism' (1983); J. G. Murphy, 'Does Kant Have a Theory of Punishment?' (1987); B. S. Byrd, 'Kant's Theory of Punishment' (1989).
2. Herbert Morris initially advocated the benefits-and-burdens theory, Morris, 'Persons and Punishment' (1968), but subsequently moved away from it, Morris, 'A Paternalistic Theory of Punishment' (1981). Jeffrie Murphy also espoused the benefits-and-burdens view, J. G. Murphy, *Retribution, Justice, and Therapy* (1979), 82–115; however, he, too, subsequently criticized the theory, see J. Murphy, 'Retributivism, Moral Education, and the Liberal State' (1985).
3. For citations in footnote: W. Sadurski, *Giving Desert Its Due* (1985), ch. 8; G. Sher, *Desert* (1987), ch. 5; J. Finnis, *Natural Law and Natural Rights* (1980), 263–64; A. Gewirth, *Reason and Morality* (1978), 294–8.
4. For critiques of the unfair-advantage theory, see, e.g. A. von Hirsch, *Past or Future Crimes* (1985), ch. 5; R. A. Duff, *Trials and Punishments* (1986), ch. 8; H. Bedau, 'Retribution and the Theory of Punishment' (1978); R. Burgh, 'Do the Guilty Deserve Punishment?' (1982). For citations in footnote, A. von Hirsch, *Doing Justice* (1976), ch. 6; von Hirsch, above n. 4; ch. 5.
5. Duff, above n. 4, 211–16.
6. For references in footnote: Sadurski, above n. 3, 229; M. Davis, 'How to Make the Punishment Fit the Crime' (1983); A. von Hirsch, 'Proportionality in the Philosophy of Punishment: From "Why Punish?" to "How Much?" ' (1990), 265–8. For further criticism of Davis's auction model, see D. Scheid, 'Davis and the Unfair-Advantage Theory of Punishment' (1990).
7. R. Wasserstrom, 'Punishment' (1980).
8. For further discussion of the censuring character of punishment, and a response to some objections by Michael Davis, see von Hirsch, above n. 6, 270–1.

9. P. F. Strawson, 'Freedom and Resentment' (1974).
10. Joel Feinberg speaks of punishment's function in recognizing the wrongfulness of the conduct in his 'Expressive Function of Punishment' (1970). Uma Narayan points out, however, that censure not only recognizes that the conduct is wrong, but confronts the actor as the agent responsible for the wrongdoing. See U. Narayan, 'Adequate Responses and Preventive Benefits' (1993).
11. Duff, above n. 4, ch. 9.
12. For discussion of the defiant-actor case, contrast Duff, above n. 4, 266, with I. Primoratz, 'Punishment as Language' (1989), 195.
13. For citations in footnote: see, H. Morris, 'A Paternalistic Theory of Punishment' (1981); J. Hampton, 'The Moral Education Theory of Punishment' (1989).
14. N. Walker, *Why Punish?* (1991), 81–2.
15. J. Kleinig, 'Punishment and Moral Seriousness' (1992); Primoratz, above n. 12, 198–202.
16. See N. Jareborg, *Essays in Criminal Law* (1988), 76–8.
17. See von Hirsch, above n. 4, ch. 5.
18. For citations in footnote: J. Braithwaite and P. Pettit, *Not Just Deserts* (1990).
19. For citations in footnote: for a sample of the views of the European theorists, see, K. Mäkelä, 'Om straffets verkningar' (1975), discussed in von Hirsch, above n. 4, 48–51. See also A. C. Ewing, *The Morality of Punishment* (1929), 94–100.
20. For reference in footnote: D. Dolinko, 'Three Mistakes of Retributivism' (1992), 1625; von Hirsch, above n. 4, 35–6.
21. For a comparable suggestion, see D. Wood, 'Dangerous Offenders and the Morality of Protective Sentencing' (1988).

CHAPTER 3

1. For the problems of the traditional utilitarian model, see A. von Hirsch and A. Ashworth (eds.), *Principled Sentencing* (1992), ch. 2 and especially 55–6.
2. J. Braithwaite and P. Pettit, *Not Just Deserts* (1990) [hereinafter 'Braithwaite and Pettit'], 64–5, 85.
3. Ibid. 78–9, 138, 142.
4. Ibid. 125–6.
5. Ibid. 101–2, 126.
6. J. Braithwaite, *Crime, Shame, and Reintegration* (1989).
7. See discussion at notes 15–20 of this chapter.
8. Braithwaite and Pettit, 142. For the discussion of the 'decremental strategy', ibid. 140–3.
9. Ibid. 153.
10. J. Feinberg, *Harm to Self* (1986), 77.
11. Braithwaite and Pettit, 101–2.
12. For citations in footnote: The quotation is from ibid. 151. For discussion of the problems of measuring deterrent effects, see von Hirsch and Ashworth, above n. 1, ch. 2.
13. R. Dahrendorf, *Law and Order* (1985).

14. See e.g. A. von Hirsch, *Doing Justice* (1976), 135–6.
15. For a critical discussion of this notion, see A. Ashworth, *Sentencing and Penal Policy* (1983), 299–305.
16. Braithwaite and Pettit, 88–90.
17. Ibid. 161–2. See also Braithwaite, above n. 6.
18. See also Ch. 2 above.
19. Braithwaite and Pettit, 161.
20. Ibid. 128.
21. See Ch. 2 above.
22. Braithwaite and Pettit, 143.
23. B. Williams, *Ethics and the Limits of Philosophy* (1985); R. Rorty, *Contingency, Irony, and Solidarity* (1989).

CHAPTER 4

1. A. von Hirsch, *Past or Future Crimes* (1985), 64–6.
2. The classic studies are T. Sellin and M. Wolfgang, *The Measurement of Delinquency* (1964); and R. F. Sparks, H. Genn, and D. Dodd, *Surveying Victims* (1977), ch. 7. For further references see A. Ashworth, *Sentencing and Criminal Justice* (1992), 84.
3. For the Minnesota Commission's rating procedures, see von Hirsch, K. Knapp, and M. Tonry, *The Sentencing Commission and its Guidelines* (1987), ch. 5.
4. Von Hirsch, above n. 1, 64–5.
5. A. Ashworth, above n. 2, 113–7.
6. See M. Wasik, 'Excuses at the Sentencing Stage' (1983).
7. Von Hirsch, above n. 1, 66–71.
8. J. Feinberg, *Harm to Others* (1984), 37–45, 55–61, 206–14.
9. A. von Hirsch and N. Jareborg, 'Gauging Criminal Harm: A Living-Standard Analysis' (1991).
10. A. Sen, *The Standard of Living* (1987).
11. Von Hirsch and Jareborg, above n. 9. For an attempt to apply this analysis in the English sentencing context, see Ashworth, above n. 2, ch. 4.
12. Von Hirsch, Knapp, and Tonry, above n. 3, 99–101.
13. For fuller discussion on how the Commission might proceed, see von Hirsch and Jareborg, above n. 9, 35–7.
14. For citations to these studies, see A. von Hirsch, M. Wasik, and J. Greene, 'Punishments in the Community and the Principles of Desert' (1989), n. 24 at 606.
15. I have changed my mind on this question, having previously supported the subjective view of severity. See, ibid, 606–9.
16. N. Walker, *Why Punish?* (1991), 99.

CHAPTER 5

1. A. von Hirsch, *Past or Future Crimes* (1985), 44, 93.
2. N. Jareborg, 'Rättvisa och repressionsnivå' (1988).

3. N. Walker, *Why Punish?* (1991), 102–3. Walker's one concession, and hardly one at that, is that if the public expects a degree of proportionality of sanctions, that would constitute a (utilitarian) reason for observing some degree of proportionality.
4. J. Kleinig, *Punishment and Desert* (1973), ch. 7.
5. Von Hirsch, above n. 1, ch. 8.
6. A. von Hirsch, K. Knapp, and M. Tonry, *The Sentencing Commission and Its Guidelines* (1987), ch. 5.
7. J. Braithwaite and P. Pettit, *Not Just Deserts* (1990), 150.
8. See Ashworth's and my critique of Braithwaite and Pettit, in Ch. 3 above.
9. A. von Hirsch and A. Ashworth (eds.), *Principled Sentencing* (1992), 52–61.
10. See e.g. National Academy of Sciences, Panel on Research on Deterrent and Incapacitative Effects, *Report* (1978).
11. For fuller discussion of the lack of impact of increased imprisonment levels on crime rates in the U.S., see National Academy of Sciences, Panel on Understanding and Control of Violent Behavior, *Report* (1993), 6–7, 292–294.
12. National Academy of Sciences, above n. 10.
13. Such a model is proposed, for example, in R. Shinnar and S. Shinnar, 'The Effect of Criminal Justice on the Control of Crime' (1975).
14. A. von Hirsch, *Doing Justice* (1976), ch. 16.
15. For discussion in footnote: in Sweden, only 1.7% of prison sentences were for durations in excess of four years. For homicide, the usual sentence is of about ten years' duration, with release on parole ordinarily occurring in the fifth year. (For citations to the Swedish imprisonment statistics, see N. Jarebourg, 'The Swedish Sentencing Law' (1933), 30.

CHAPTER 6

1. See P. Robinson, 'Hybrid Principles for the Distribution of Criminal Sanctions' (1987).
2. Ibid. See also A. von Hirsch, 'Hybrid Principles in Allocating Sanctions: A Reply to Professor Robinson' (1987).
3. These examples are taken from von Hirsch, above n. 2.
4. Swedish Criminal Code, chs. 29 and 30. For commentary on and translation of these provisions, see A. von Hirsch and A. Ashworth (eds.), *Principled Sentencing* (1992), 292–307.
5. Ibid. 299.
6. See P. W. Greenwood, *Selective Incapacitation* (1982).
7. Greenwood claimed that up to a 20% reduction in robberies could be achieved, ibid.; see also J. Q. Wilson, *Thinking About Crime* revised edn. (1983), ch. 8. For a critique of these effectiveness claims, see A. von Hirsch, *Past or Future Crimes* (1985), chs. 10–11; National Academy of Sciences, Panel on Research on Criminal Careers, *Report* (1986); A. von Hirsch, 'Selective Incapacitation Reexamined' (1988).
8. See F. Schoeman, 'On Incapacitating the Dangerous' (1979).
9. Using Dworkin's model has been suggested in A. E. Bottoms and R. Brownsword, 'Dangerousness and Rights' (1983).

10. R. Dworkin, *Taking Rights Seriously* (1977), ch. 7.
11. Ibid. 200.
12. See e.g. J. Raz, *The Morality of Freedom* (1986), chs. 7 and 8; see also N. MacCormick, *Legal Reasoning and Legal Theory* (1978).
13. See J. Reiman's views in J. Reiman and E. van den Haag, 'On the Common Saying that It Is Better that Ten Guilty Persons Escape than One Innocent Person Suffer' (1990).
14. See Robinson, above n. 1, 39.
15. For discussion in footnote: such a period of civil confinement has been suggested in D. Wood, 'Dangerous Offenders and the Morality of Protective Sentencing' (1988).
16. See A. L. Ross, *Deterring the Drinking Driver* (1982), 60–9.
17. See the Swedish Government's report introducing proposed amendments of the drinking-and-driving law, Prop. 1989/90:2, 38–9.
18. Bottoms and Brownsword, above n. 9. For reference in footnote ibid., Dworkin, above n. 10, 199–200.
19. See National Academy of Sciences, Panel on Research on Criminal Careers, above n. 7; von Hirsch, 'Selective Incapacitation Reexamined', above n. 7.
20. For references in footnote: see Criminal Justice Act 1991, ss. 1(2)(b), 2(2)(b), 31. For commentary on these provisions, see A. Ashworth, *Sentencing and Criminal Justice* (1992), 163–7.
21. N. Morris, *Madness and the Criminal Law* (1982), ch. 5.
22. See H. J. Bruns, *Das Recht der Strafzumessung*, 2nd edn. (1985), 105–9. For a critique of the 'Spielraumtheorie', see A. von Hirsch and N. Jareborg, *Strafmass und Strafgerechtigkeit* (1991), 23–7.
23. Morris, above n. 21, 198–9.
24. For fuller discussion see von Hirsch, *Past or Future Crimes*, above n. 7, ch. 4; and von Hirsch, 'Selective Incapacitation Revisited', above n. 7.
25. Morris, above n. 21, 202–5.
26. For a critique of the indeterminacy of Morris's view, see Ch. 7 below; von Hirsch, *Past or Future Crimes*, ch. 12. For the similar indeterminacy of the 'Spielraumtheorie', see von Hirsch and Jareborg, above n. 22, 23–7.
27. See N. Morris and M. Miller, 'Predictions of Dangerousness' (1985).
28. See discussion of 'selective incapacitation' in text accompanying notes 6 and 7 of this chapter.
29. N. Morris and M. Tonry, *Between Prison and Probation* (1990), 104–8.
30. For discussion of such issues of 'parsimony for whom?' see A. von Hirsch, 'Equality, "Anisonomy", and Justice' (1984), 1105–7.

CHAPTER 7

1. See e.g. J. Petersilia, *Expanding Options for Criminal Sentencing* (1987); M. Tonry and R. Will, *Intermediate Sanctions* (1989); N. Morris and M. Tonry, *Between Prison and Probation* (1990); A. von Hirsch and A. Ashworth (eds.), *Principled Sentencing* (1992), ch. 6; J. M. Byrne, A. J. Lurigio, and J. Petersilia (eds.), *Smart Sentencing: The Emergence of Intermediate Sanctions* (1992).
2. Tonry and Will, above n. 1; A. von Hirsch and A. Ashworth, above n. 1, ch. 6.

3. For reference in footnote, see e.g. A. von Hirsch, K. Knapp, and M. Tonry, *The Sentencing Commission and its Guidelines* (1987), ch. 5.
4. Criminal Justice Act 1991, ss. 6(2), 18(2). See also M. Wasik and R. Taylor, *Blackstone's Guide to the Criminal Justice Act of 1991* (1991), chs. 2 and 3; A. Ashworth, *Sentencing and Criminal Justice* (1992), ch. 10.
5. A. von Hirsch, M. Wasik, and J. Greene, 'Punishments in the Community and the Principles of Desert' (1989). An earlier version appears as M. Wasik and A. von Hirsch, 'Non-custodial Penalties and the Principles of Desert' (1988).
6. Morris and Tonry, above n. 1.
7. Von Hirsch, Wasik, and Greene, above n. 5, 600–6.
8. Ibid., n. 18 at 599, 615–16.
9. F. E. Zimring and G. Hawkins, *The Scale of Imprisonment* (1991), 174. In California, large numbers of offenders have their paroles revoked for failing drug tests.
10. Von Hirsch, Wasik, and Greene, above n. 5, 599–600.
11. See e.g. A. von Hirsch, *Past or Future Crimes* (1985), ch. 11; A. von Hirsch, 'Selective Incapacitation Reexamined' (1988).
12. A. E. Bottoms, 'The Concept of Intermediate Sanctions and Its Relevance for the Probation Service' (1989), 91.
13. See von Hirsch, Knapp, and Tonry, above n. 3, ch. 1.
14. Morris and Tonry, above n. 1, 51–4.
15. Von Hirsch, Wasik, and Greene, above n. 5, 605.
16. Ibid. 600–2.
17. Ibid. 602–4. See also, P. Robinson, 'A Sentencing System for the 21st Century?' (1987).
18. Von Hirsch, Wasik, and Greene, above n. 5, 604–6.
19. Ashworth, above n. 4, 246.
20. Von Hirsch, Wasik, and Greene, above n. 5, 610.
21. Morris and Tonry, above n. 1.
22. Ibid. 93.
23. Ibid. 103–4.
24. Ibid. 79.
25. Ibid. 101.
26. Ibid. 48–51.
27. Ibid. 105–8.
28. Ibid. 104.
29. Ibid. 77–8, 90–1.
30. See T. Clear and P. Hardyman, 'The New Intensive Supervision Movement' (1990), 48–53.
31. Morris and Tonry, above n. 1, 77–8.

CHAPTER 8

1. See A. von Hirsch and A. Ashworth, *Principled Sentencing* (1992), ch. 1; and particularly 40–50.
2. Ibid. ch. 5.
3. J. Braithwaite, *Crime, Shame, and Reintegration* (1989).

4. R. A. Duff, 'Punishment, Expression and Penance' (1991).
5. Ibid. 241.
6. Ibid. 242–3; R. A. Duff, *Trials and Punishments* (1986), 278–80.
7. Duff, *Trials and Punishments*, 91–3.
8. Duff, above n. 4.
9. Ibid. 242.
10. Ibid., 244.
11. See Ch. 7 above.
12. Duff, above n. 4, 245.
13. A. von Hirsch, 'Punishment to Fit the Criminal' (1988).
14. Von Hirsch and Ashworth, above n. 1, ch. 5.
15. For more on the idea of appreciation of consequences, see D. Husak and A. von Hirsch, 'Culpability and Mistake of Law' (1993).
16. See A. von Hirsch, K. Knapp, and M. Tonry, *The Sentencing Commission and its Guidelines* (1987), chs. 1, 3.
17. For discussion of the Minnesota guidelines, see ibid., ch. 5.
18. For discussion of the Finnish statute, see T. Lappi-Seppälä, 'Penal Policy and Sentencing Theory in Finland' (1992); for discussion of the Swedish and English statutes, see von Hirsch and Ashworth, above n. 1, 282–307. For more on the English approach, see A. Ashworth, *Sentencing and Criminal Justice* (1992).

CHAPTER 9

1. See e.g. Rummel *v*. Estelle, 445 US 263 (1980); Solem *v*. Helm, 463 US 277 (1983).
2. J. Murphy, 'Cruel and Unusual Punishments' (1979), 223.
3. Ibid. 237.
4. D. L. MacKenzie and D. Parent, 'Boot Camp Prisons for Young Offenders' (1992).
5. For a description of this project, see D. C. McDonald, *Prisons without Walls* (1986).

CHAPTER 10

1. National Council on Crime and Delinquency, Council of Judges, *Model Sentencing Act*, 2nd edn. (1972). The first edition of the proposed Act was published in 1963.
2. Model Sentencing Act, ss. 1, 5, 9.
3. See e.g. K. G. Armstrong, 'The Retributivist Hits Back' (1961); H. J. McCloskey 'A Non-Utilitarian Approach to Punishment' (1965); and particularly, J. Kleinig, *Punishment and Desert* (1973).
4. A. von Hirsch, *Doing Justice* (1976).
5. For references in footnote: A. Ashworth, 'Criminal Justice and Deserved Sentences' (1989); A. E. Bottoms, 'The Concept of Intermediate Sanctions and Its Relevance for the Probation Service' (1989); R. A. Duff, *Trials and Punishments* (1986), ch. 9; N. Jareborg, *Essays in Criminal Law* (1988), ch. 5;

N. Jareborg, 'Rättvisa Och Repressionsnivå' (1988); R. Singer, *Just Deserts* (1979); M. Wasik, 'Guidance, Guidelines and Criminal Record' (1987); A. von Hirsch, *Past or Future Crimes* (1985); A. von Hirsch and N. Jareborg, *Strafmass und Strafgerechtigkeit* (1991).

6. See e.g. J. Q. Wilson, *Thinking about Crime*, revised edn. (1983); E. van den Haag, 'Punishment: Desert and Control' (1987).

7. J. Reiman and S. Headlee, 'Marxism and Criminal Justice Policy' (1981).

8. K. Stenson, 'Making Sense of Crime Control' (1991), 22.

9. B. Hudson, *Justice Through Punishment* (1987), 57–8.

10. Wilson, above n. 6, ch. 8.

11. Compare E. van den Haag, *Punishing Criminals* (1975) with van den Haag, above n. 6.

12. Van den Haag, above n. 6, 1255–6.

13. Wilson, above n. 6, ch. 8.

14. See Office of National Drug Control Policy, *National Drug Control Strategy* (Washington, DC, 1989).

15. For an analysis of the structure of the Minnesota guidelines, see A. von Hirsch, K. Knapp, and M. Tonry, *The Sentencing Commission and Its Guidelines* (1987), ch. 5.

16. US Sentencing Commission, Supplementary Report on the Initial Guidelines and Policy Statements (1987), 60–4. For a critical analysis of the Federal guidelines see A. von Hirsch, 'Federal Sentencing Guidelines' (1989).

17. I. Nagel, 'Structuring Sentencing Discretion' (1990), 916–25.

18. For an analysis of the Finnish law, see T. Lappi-Seppälä, 'Penal Policy and Sentencing Theory in Finland' (1992).

19. The statute is set forth in Finnish Penal Code, ch. 6.

20. For citation in footnote, Swedish Penal Code, former s. 1:7.

21. Swedish Penal Code, chs. 29 and 30. The text is set forth in A. von Hirsch and A. Ashworth (eds.), *Principled Sentencing* (1992), 302–7. For an analysis of the law, see ibid. 292–300.

22. Hudson, above n. 9, ch. 3.

23. See J. Cohen and M. Tonry, 'Sentencing Reforms and Their Impacts' (1983), 353–411.

24. Ibid. 383.

25. For fuller analysis, see A. von Hirsch and J. Mueller, 'California's Determinate Sentencing Law' (1984), 267–70. California has enacted some mandatory minima requiring commitment in certain cases, but no systematic guidance limiting prison commitments to more serious crimes.

26. Von Hirsch, Knapp, and Tonry, above n. 15, 76–83.

27. Ibid.; see also New York State Committee on Sentencing Guidelines, *Determinate Sentencing: Report and Recommendations* (1985).

28. See citations in n. 16 above.

29. For an analysis of those guidelines, see von Hirsch, Knapp, and Tonry, above n. 15, ch. 5.

30. Ibid., chs. 5 and 8.

31. The changes were made at the end of 1989, and are set forth in *Minnesota Sentencing Guidelines Annotated* (1990), Pts IIB and IV.

32. See F. A. Allen, *The Decline of the Rehabilitative Ideal* (1981).

33. On the limited effectiveness of currently-available incapacitative techniques for reducing crime rates, for example, see National Academy of Sciences, Panel on Research on Criminal Careers, *Report* (1986); A. von Hirsch, 'Selective Incapacitation Reexamined' (1988).

34. Von Hirsch, Knapp, and Tonry, above n. 15, 88–9.

35. See P. W. Greenwood, *Selective Incapacitation* (1982) for a description of the strategy. For critiques, see citations at n. 33, above.

36. Von Hirsch, above n. 5, ch. 11; von Hirsch, above n. 33.

37. E. Currie, *Confronting Crime* (1985), 81–101.

38. For references in footnote: F. T. Cullen and K. E. Gilbert, *Reaffirming Rehabilitation* (1982); Hudson, above n. 9, 170–6. But see T. Mathiesen, *Prison on Trial* (1990), ch. 2.

39. See A. von Hirsch and L. Maher, 'Should Penal Rehabilitationism be Revived?' (1992).

40. D. Anglin and Y. Hser, 'The Treatment of Drug Abuse' (1990).

41. Von Hirsch and Maher, above n. 39.

42. Hudson, above n. 9, 164–9.

43. J. Q. Wilson, *Thinking About Crime* (1975), 42–57.

44. Von Hirsch, above n. 4, 147–8.

45. Hudson, above n. 9, ch. 4.

46. M. Gardner, 'The Renaissance of Retribution' (1976), 804–5.

47. See K. G. Lång, 'Upplever fängelsestraffetet en renässans?' (1989); T. Lappi-Seppälä, above n. 18, 116–18.

48. Police and Criminal Evidence Act 1984.

49. UK Government White Paper, *Crime, Justice and Protecting the Public* (1990). For commentary on the White Paper, see M. Wasik and A. von Hirsch, 'Statutory Sentencing Principles' (1990).

50. Increases in prison terms will require costly added prison construction, unless present policies of single-celling prisoners were to be altered.

51. See Criminal Justice Act 1991, s. 1(2)(a); A. Ashworth, *Sentencing and Criminal Justice* (1992), 228–31.

52. Criminal Justice Act 1991, s. 29; Ashworth, above n. 51, 151–2.

53. Criminal Justice Act 1991, s. 33; M. Wasik and R. D. Taylor, *Blackstone's Guide to the Criminal Justice Act of 1991* (1991), ch. 5.

54. Fängelsestraffkommittén, *Påföljd för Brott* (1986).

55. See J. Graham, 'Decarceration in the Federal Republic of Germany' (1990).

EPILOGUE

1. M. Tonry, 'Proportionality, Interchangeability, and Intermediate Punishments' (1992).

2. A. von Hirsch, K. Knapp, and M. Tonry, *The Sentencing Commission and Its Guidelines* (1987), 54–6, 97–9.

3. A. von Hirsch, *Doing Justice* (1976), ch. 17.

4. R. F. Sparks, 'A Critique of Marxist Criminology' (1980), 159–210.

5. J. Lea and J. Young, *What Is To Be Done About Law and Order?* (1984).

6. Sparks, above n. 4.

7. D. Garland, *Punishment and Modern Society* (1991), ch. 5.
8. Tonry, above n. 1.
9. Ibid.
10. The text of the Committee's proposals are set forth in A. von Hirsch, 'Principles for Choosing Sanctions' (1987), 191–5, and this particular provision is set forth at ibid. 193. The statutory version of the provision is different and somewhat broader. See Swedish Penal Code, s. 30:9. (The text of the law is set out in A. von Hirsch and A. Ashworth (eds.), *Principled Sentencing* (1992), 302–7.)
11. In the footnote: For more on sentencing factors and personal autonomy, see A. von Hirsch, 'Desert and Previous Convictions in Sentencing' (1981), 607–11.
12. M. Tonry, 'Salvaging the Sentencing Guidelines in Seven Easy Steps' (1992), especially 356.
13. 18 USC s. 994(j).
14. For reference in footnote, see von Hirsch, Knapp, and Tonry, above n. 2, 92–3.

Bibliography

ADLER, J., *The Urgings of Conscience: A Theory of Punishment* (Philadelphia, 1991).

ALLEN, F. A., *The Decline of the Rehabilitative Ideal: Penal Policy and Social Purpose* (New Haven, Conn., 1981).

American Friends Service Committee, *Struggle for Justice* (New York, 1971).

ANGLIN, D. M., and HSER, Y., 'The Treatment of Drug Abuse', in Tonry, M., and Wilson, J. Q. (eds.), *Drugs and Crime* (Chicago, 1990), 393–460.

ARMSTRONG, K. G., 'The Retributivist Hits Back', *Mind* 70 (1961), 471–91.

ASHWORTH, A., *Sentencing and Penal Policy* (London, 1983).

—— 'Criminal Justice and Deserved Sentences', *Criminal Law Review* (1989), 340–55.

—— 'Non-Custodial Sentences', *Criminal Law Review* (1992), 242–51.

—— *Sentencing and Criminal Justice* (London, 1992).

BEDAU, H. A. 'Retribution and the Theory of Punishment', *Journal of Philosophy* 75 (1978), 601–20.

—— 'Classification-Based Sentencing: Some Conceptual and Ethical Problems', *New England Journal on Criminal and Civil Confinement* 10 (1984), 1–26.

BENTHAM, J., *Introduction to the Principles of Morals and Legislation* (London, 1982) (first published 1780).

BLUMSTEIN, A., COHEN, J., MARTIN, S. E., and TONRY, M., (eds.), *Research on Sentencing: The Search for Reform* (Washington, DC, 1983).

BOTTOMS, A. E., 'The Concept of Intermediate Sanctions and Its Relevance for the Probation Service', in Shaw, R., and Haines, K. (eds.), *The Criminal Justice System: A Central Role for the Probation Service* (Cambridge, 1989), 84–102.

—— and BROWNSWORD, R., 'Dangerousness and Rights', in Hinton, J. W. (ed.), *Dangerousness: Problems of Assessment and Prediction* (London, 1983), 9–22.

BRAITHWAITE, J., *Crime, Shame, and Reintegration* (Cambridge, 1989).

—— and PETTIT, P., *Not Just Deserts: A Republican Theory of Justice* (Oxford, 1990).

BRUNS, H. J., *Das Recht des Strafzumessung* (Cologne, 1985).

BURGH, R., 'Do the Guilty Deserve Punishment?' *Journal of Philosophy* 79 (1982), 193–210.

BYRD, B. S., 'Kant's Theory of Punishment: Deterrence in Its Threat, Retribution in Its Execution', *Law and Philosophy* 8 (1989), 151–200.

BYRNE, J. M., LURIGIO, A. J., and PETERSILIA, J. (eds.), *Smart Sentencing: The Emergence of Intermediate Sanctions* (Newbury Park, Calif., 1992).

CLEAR, T., and HARDYMAN, P., 'The New Intensive Supervision Movement', *Crime and Delinquency* 36 (1990), 42–60.

COHEN, J., and TONRY, M., 'Sentencing Reforms and Their Impacts', in Blumstein, Cohen, Martin, and Tonry (eds.), *Research on Sentencing: The Search for Reform*, above, 305–459.

CULLEN, F. T., and GILBERT, K. E., *Reaffirming Rehabilitation* (Cincinnati, 1982).

CURRIE, E., *Confronting Crime: An American Challenge* (New York, 1985).

DAHRENDORF, R., *Law and Order* (London, 1985).

DAVIS, M., 'How to Make the Punishment Fit the Crime', *Ethics* 93 (1983), 726–52.
—— 'Punishment as Language: Misleading Analogy for Desert Theorists', *Law and Philosophy* 10 (1991), 311–22.
DOLINKO, D., 'Some Thoughts About Retributivism', *Ethics* 101 (1991), 537–59.
—— 'Three Mistakes of Retributivism', *UCLA Law Review* 39 (1992), 1623–57.
DUFF, R. A., *Trials and Punishments* (Cambridge, 1986).
—— 'Punishment and Penance: A Reply to Harrison', *Aristotelian Society—Supplementary Volumes* 62 (1988), 153–67.
—— 'Punishment, Expression, and Penance', in Jung, H., and Müller-Dietz, H. (eds.), *Recht und Moral: Beiträge zu einer Standortsbestimmung* (Baden-Baden, 1991).
DWORKIN, R., *Taking Rights Seriously* (Cambridge, Mass., 1977).
EWING, A. C., *The Morality of Punishment* (Montclair, NJ, 1970) (first published 1929).
Fängelsestraffkommittén, *Päföljd för Brott*, Stadens Offentiga Utredningar (1986), vols. 13–15.
FEINBERG, J., 'The Expressive Functions of Punishment', in *Doing and Deserving: Essays in the Theory of Responsibility*, ed. Feinberg, J. (Princeton, NJ, 1970), 53–69.
—— *Harm to Others* (Oxford, 1984).
—— *Harm to Self* (Oxford, 1986).
FINNIS, J., *Natural Law and Natural Rights* (Oxford, 1980).
FLETCHER, G., *Rethinking Criminal Law* (1978).
FLOUD, J., and YOUNG, W., *Dangerousness and Criminal Justice* (London, 1981).
FOGEL, D., *We Are the Living Proof . . .* (Cincinnati, 1975).
FRANKEL, M. E., *Criminal Sentences: Law Without Order* (New York, 1972).
GARDNER, M., 'The Renaissance of Retribution: An Examination of "Doing Justice" ', *Wisconsin Law Review* (1976), 781–815.
GARLAND, D., *Punishment and Modern Society* (Oxford, 1991).
GEWIRTH, A., *Reason and Morality* (Chicago, 1978).
GOTTFREDSON, D. M., WILKINS, L., and HOFFMAN, P., *Guidelines for Parole and Sentencing: A Policy Control Method* (Lexington, Mass., 1978).
GRAHAM, J., 'Decarceration in the Federal Republic of Germany', *British Journal of Criminology*, 30 (1990), 150–70.
GREENBERG, D., and HUMPHRIES, D., 'Economic Crisis and the Justice Model: A Skeptical View', *Crime and Delinquency* 28 (1982), 601–17.
GREENE, J., 'Structuring Criminal Fines: Making an "Intermediate" Penalty More Useful and Equitable', *Justice Systems Journal* 13 (1988), 37–50.
GREENWOOD, P. W., *Selective Incapacitation* (Santa Monica, Calif., 1982).
GRYGIER, T., 'A Canadian Approach or American Band-Wagon?' *Canadian Journal of Criminology* 30 (1988), 165–72.
HAMPTON, J., 'The Moral Education Theory of Punishment', *Philosophy and Public Affairs* 13 (1989), 208–38.
HART, H. L. A., *Punishment and Responsibility: Essays in the Philosophy of Law* (Oxford, 1968).
HUDSON, B., *Justice Through Punishment: A Critique of the 'Justice Model' of Corrections* (London, 1987).
HUSAK, D., *Philosophy of Criminal Law* (Totowa, NJ, 1987).
—— 'Why Punish the Deserving?' *Nous* 26 (1992), 447–464.

—— *Drugs and Rights* (Cambridge, 1992).

—— and VON HIRSCH, A., 'Culpability and Mistake of Law', in Shute, S., Gardner, J., and Horder, J. (eds.), *Action and Value in the Criminal Law* (Oxford, 1993) (forthcoming).

JAREBORG, N., *Essays in Criminal Law* (Uppsala, 1988).

—— 'Rättvisa och repressionsnivå', in *Rättsdogmatikens alternativ*, ed. Tuori, K. (Helsinki, 1988).

—— 'The Swedish Sentencing Reform', Paper presented at Colston International Sentencing Symposium, Law Faculty, University of Bristol, April 1993.

KLEINIG, J., *Punishment and Desert* (The Hague, 1973).

—— 'Punishment and Moral Seriousness', *Israel Law Review* 25 (1992), 401–21.

LACEY, N., *State Punishment* (London, 1987).

LÅNG, K. G., 'Upplever fängelsestraffet en renässans?' *Nordisk Tidsskrift for Kriminalvidenskab* 76 (1989), 83–94.

LAPPI-SEPPÄLÄ, T., 'Penal Policy and Sentencing Theory in Finland', *Canadian Journal of Law and Jurisprudence* 5 (1992), 95–120.

LEA, J., and YOUNG, J., *What Is To Be Done About Law and Order?* (Harmondsworth, 1984).

McCloskey, H. J., 'A Non-Utilitarian Approach to Punishment,' *Inquiry* 8 (1965), 249–63.

MACCORMICK, N., *Legal Reasoning and Legal Theory* (Oxford, 1978).

—— *Legal Right and Social Democracy* (Oxford, 1982).

McDONALD, D. C., *Prisons Without Walls: Community Service Sentences in New York City* (New Brunswick, NJ, 1986).

MACKENZIE, D. L, and PARENT, D., 'Boot Camp Prisons for Young Offenders', in Byrne, Lurigio, and Petersilia (eds.), *Smart Sentencing* above.

MACKIE, J. L., *Persons and Values* (Oxford, 1985).

MAKELA, K., 'Om straffets verkningar', *Jurisprudentia* 6 (1975), 237–80.

MATHIESEN, P., *Prison on Trial* (London, 1990).

Minnesota Sentencing Guidelines Annotated (St Paul, Minn., 1992).

MORRIS, H., 'Persons and Punishment', *The Monist* 52 (1968), 475–501.

—— 'A Paternalistic Theory of Punishment', *American Philosophical Quarterly* 18 (1981), 263–71.

MORRIS, N., *Madness and the Criminal Law* (Chicago, 1982).

—— and MILLER M., 'Predictions of Dangerousness', in *Crime and Justice: An Annual Review of Research*, ed. Tonry, M., and Morris, N., 6 (1985), 1–50.

—— and TONRY, M., *Between Prison and Probation: Intermediate Punishments in a Rational Sentencing System* (New York, 1990).

MURPHY, J. G., 'Marxism and Retribution', *Philosophy and Public Affairs* 2 (1973), 217–43.

—— 'Cruel and Unusual Punishments', in *Retribution, Justice, and Therapy* ed. Murphy, J. G. (Dordrecht, 1979).

—— 'Retributivism, Moral Education, and the Liberal State', *Criminal Justice Ethics* 4(1) (1985), 3–11.

—— 'Does Kant Have a Theory of Punishment?' *Columbia Law Review* 87 (1987), 509–32.

NAGEL, I., 'Structuring Sentencing Discretion: The New Federal Sentencing Guidelines', *Journal of Criminal Law and Criminology* 80 (1990), 883–943.

NARAYAN, U., 'Moral Education and Criminal Punishment', Paper presented at 21st Conference on Value Inquiry, Drew University, NJ, April, 1993.

—— 'Adequate Responses and Preventive Benefits: Justifying Censure and Hard Treatment in Legal Punishment', *Oxford Journal of Legal Studies* 13 (1993), (forthcoming).

National Academy of Sciences, Panel on Research on Deterrent and Incapacitative Effects, *Report*, in Blumstein, A., Cohen, J., and Nagin, D. (eds.), *Deterrence and Incapacitation: Estimating the Effects of Criminal Sanctions on Crime Rates* (Washington, DC, 1978), 1–90.

—— Panel on Research on Criminal Careers, *Report*, in Blumstein, A., Cohen, J., Roth, J. A., and Visher, C. (eds.), *Criminal Careers and 'Career Criminals'*, (Washington, DC, 1986), 1.

—— Panel on Understanding and Control of Violent Behavior, *Report*. In Reiss, A. J. Jr., and Roth, J. A. (eds.), *Understanding and Preventing Violence*, (Washington, D.C., 1993).

National Council on Crime and Delinquency, Council of Judges, *Model Sentencing Act*, 2nd. edn. (Hackensack, NJ, 1972). [Portions of this proposed statute are reprinted in von Hirsch and Ashworth (eds.), *Principled Sentencing*, below.]

New York State Committee on Sentencing Guidelines, *Determinate Sentencing: Report and Recommendations* (New York, 1985).

Office of National Drug Control Policy, *National Drug Policy* (Washington, DC, 1989).

PALMER, T., *The Re-Emergence of Correctional Intervention*, (Newbury Park, Calif., 1992).

PETERSILIA, J., *Expanding Options for Criminal Sentencing* (Santa Monica, Calif., 1987).

—— and TURNER, S., 'An Evaluation of Intensive Probation in California', *Journal of Criminal Law and Criminology* 82 (1991), 610–58.

PRIMORATZ, I., 'Punishment as Language', *Philosophy* 64 (1989), 187–205.

RAZ, J., *The Morality of Freedom* (Oxford, 1986).

REIMAN, J., 'Marxist Explanations and Radical Misinterpretations: A Reply to Greenberg and Humphries', *Crime and Delinquency* 28 (1982), 610–17.

—— and HEADLEE, S., 'Marxism and Criminal Justice Policy', *Crime and Delinquency* 27 (1981), 24–47.

—— and VAN DEN HAAG, E., 'On the Common Saying that It Is Better that Ten Guilty Persons Escape than One Innocent Person Suffer', in Paul, E. F., Miller, F. D., Jr., and Paul, J. (eds.), *Crime, Culpability, and Remedy* (Oxford, 1990).

ROBINSON, P., 'A Sentencing System for the 21st Century?' *Texas Law Review* 66 (1987), 1–6.

—— 'Hybrid Principles for the Distribution of Criminal Sanctions', *Northwestern Law Review* 82 (1987), 19–42.

RORTY, R., *Contingency, Irony, and Solidarity* (Cambridge, 1989).

ROSS, A. L., *Deterring the Drinking Driver: Legal Policy and Social Control* (Lexington, Mass., 1982).

ROTHMAN, D., *Conscience and Convenience: The Asylum and its Alternatives in Progressive America* (Boston, 1980).

SADURSKI, W., *Giving Desert Its Due: Social Justice and Legal Theory* (Dordrecht, 1985).

SCHEID, D. E., 'Kant's Retributivism', *Ethics* 93 (1983), 262–82.

—— 'Davis and the Unfair-Advantage Theory of Punishment', *Philosophical Topics* 18 (1990), 143–170.

SCHIFF, M. F., 'Gauging the Intensity of Criminal Penalties: Developing the Criminal Penalty Severity Scale' (Ph.D. Dissertation: New York University, Robert F. Wagner Graduate School of Public Service, New York, 1992).

SCHOEMAN, F., 'On Incapacitating the Dangerous', *American Philosophical Quarterly* 16 (1979), 27–35.

SELLIN, T., and WOLFGANG, M., *The Measurement of Delinquency* (New York, 1964).

SEN, A., *The Standard of Living* (Cambridge, 1987).

SHER, G., *Desert* (Princeton, NJ, 1987).

SHINNAR, R., and SHINNAR, S., 'The Effects of Criminal Justice on the Control of Crime: A Quantitative Approach', *Law and Society Review* 9 (1975), 581–611.

SIMMONS, A. J., *Moral Principles and Political Obligation* (Princeton, NJ, 1979).

SINGER, R., *Just Deserts: Sentencing Based on Equality and Desert* (Cambridge, Mass., 1979).

SPARKS, R. F., 'A Critique of Marxist Criminology', *Crime and Justice: An Annual Review of Research*, (eds.) Morris, N., and Tonry, M., 2 (1980), 159–210.

SPARKS, R. F., GENN, H., and DODD, D., *Surveying Victims: A Study of the Measurement of Criminal Victimization* (Chichester, 1977).

STENSON, K., 'Making Sense of Crime Control', in Stenson, K., and Cowell, D. (eds.), *The Politics of Crime Control* (London, 1991).

STRAWSON, P. F., 'Freedom and Resentment'. in *Freedom and Resentment and Other Essays*, ed. Strawson, P. F. (London, 1974), 1–25.

THOMAS, D. A., *Principles of Sentencing*, 2nd edn. (London, 1979).

TONRY, M., 'Salvaging the Sentencing Guidelines in Seven Easy Steps', *Federal Sentencing Reporter* (May/June 1992), 355–9.

—— 'Proportionality, Interchangeability, and Intermediate Punishments', in Dobash, R., Duff, R. A., and Marshall, S. (eds.), *Penal Theory and Penal Practice* (Manchester, 1993) (forthcoming).

—— and WILL, R., 'Intermediate Sanctions', (Washington, DC, National Institute of Corrections, 1989).

Twentieth Century Fund, Task Force on Criminal Sentencing, *Fair and Certain Punishment* (New York, 1976).

UK Government White Paper, *Crime, Justice and Protecting the Public* (London, 1990).

USA Sentencing Commission, *Supplementary Report on the Initial Guidelines and Policy Statements* (Washington, DC, 1987).

VAN DEN HAAG, E., *Punishing Criminals: Concerning a Very Old and Painful Question* (New York, 1975).

—— 'Punishment: Desert and Control', *Michigan Law Review* 85 (1987), 1250–60.

VON HIRSCH, A., *Doing Justice: The Choice of Punishments* (New York, 1976).

—— 'Desert and Previous Convictions in Sentencing', *Minnesota Law Review* 65 (1981), 591–634.

—— 'Commensurability and Crime Prevention: Evaluating Formal Sentencing Structures and Their Rationale', *Journal of Criminal Law and Criminology* 74 (1983), 209–15.

—— 'Equality, "Anisonomy", and Justice: A Review of *Madness and the Criminal Law*', *Michigan Law Review* 82 (1984), 1093–1112.

VON HIRSCH, A., *Past or Future Crimes: Deservedness and Dangerousness in the Sentencing of Criminals* (New Brunswick, NJ, 1985).

—— 'Guidance by Numbers or Words—Numerical versus Narrative Guidelines for Sentencing', in Wasik, M., and Pease, K., (eds.), *Sentencing Reform: Guidance or Guidelines?* (Manchester, 1987).

—— 'Guiding Principles for Sentencing: The Proposed Swedish Law', *Criminal Law Review* (1987), 746–55.

—— 'Principles for Choosing Sanctions: Sweden's Proposed Sentencing Statute', *New England Journal on Criminal and Civil Confinement* 13 (1987), 171–95.

—— 'Hybrid Principles in Allocating Sanctions: A Response to Professor Robinson', *Northwestern Law Review* 82 (1987): 64–72.

—— 'Punishment to Fit the Criminal', *The Nation* (25 June, 1988), 901–2.

—— 'Selective Incapacitation Reexamined: The National Academy of Sciences' Report on Criminal Careers and "Career Criminals" ', *Criminal Justice Ethics* 7(1) (1988), 19–35.

—— 'Federal Sentencing Guidelines: Do They Provide Principled Guidance?' *American Criminal Law Review* 27 (1989), 367–90.

—— 'Proportionality in the Philosophy of Punishment: From "Why Punish?" to "How Much?" ', *Criminal Law Forum* 1 (1990), 259–90.

—— 'The Ethics of Community-Based Sanctions', *Crime and Delinquency* 36 (1990), 162–173.

—— 'Criminal Record Rides Again', *Criminal Justice Ethics* 10(2) (1991) 2, 55–7.

—— 'Scaling Intermediate Punishments: A Comparison of Two Models', in Byrne, Lurigio, and Petersilia (eds.), *Smart Sentencing* above.

—— 'Proportionality in the Philosophy of Punishment', *Crime and Justice: A Review of Research* ed. Tonry, M., (1992), 55–98.

—— and ASHWORTH, A. (eds.), *Principled Sentencing* (Edinburgh, 1993).

—— and GREENE, J., 'When Should Reformers Support Creation of Sentencing Guidelines?' *Wake Forest Law Review* 28/2 (1993) (forthcoming).

—— and HANRAHAN, K. J., *The Question of Parole: Retention, Reform or Abolition?* (Cambridge, Mass., 1979).

—— and JAREBORG, N., 'Provocation and Culpability', in Schoeman, F. (ed.), *Responsibility, Character, and the Emotions: New Essays in Moral Psychology* (Cambridge, 1987), 241–55.

—— —— 'Sweden's Sentencing Statute Enacted', *Criminal Law Review* (1989), 275–81.

—— —— 'Gauging Criminal Harm: A Living-Standard Analysis', *Oxford Journal of Legal Studies* 11 (1991), 1–38.

—— —— *Strafmass und Strafgerechtigkeit: Die deutsche Strafzumessungslehre und das Prinzip der Tatproportionalität* (Bad Godesberg, 1991).

——and MAHER, L., 'Should Penal Rehabilitation Be Revived?' *Criminal Justice Ethics* 11(1) (1992), 35–40. [This article also appears in von Hirsch and Ashworth (eds.), *Principled Sentencing*, above, 41–50.]

—— and MUELLER, J. M., 'California's Determinate Sentencing Law: An Analysis of Its Structure', *New England Journal on Criminal and Civil Confinement* 10 (1984), 253–300.

—— KNAPP., K., and TONRY, M., *The Sentencing Commission and Its Guidelines* (Boston, 1967).

—— WASIK, M., and GREENE, J., 'Punishments in the Community and the Principles of Desert', *Rutgers Law Journal* 20 (1989), 595–618. This article also appears in von Hirsch and Ashworth (eds.), *Principled Sentencing*, above, 368–88.

WALKER, N., *Why Punish?* (London, 1991).

WASIK, M., 'Excuses at the Sentencing Stage', *Criminal Law Review* (1983), 450–65.

—— 'Guidance, Guidelines and Criminal Record', in Wasik, M., and Pease, K. (eds.), *Sentencing Reform: Guidance or Guidelines?'* (Manchester, 1987), 104–5.

—— and PEASE, K. (eds.), *Sentencing Reform: Guidance or Guidelines?* (Manchester, 1987).

—— and TAYLOR, R. D., *Blackstone's Guide to the Criminal Justice Act of 1991* (London, 1991).

—— and VON HIRSCH, A., 'Non-Custodial Penalties and the Principles of Desert', *Criminal Law Review* (1988), 555–72.

—— —— 'Statutory Sentencing Principles: The 1990 White Paper', *Modern Law Review* 53 (1990), 508–17.

WASSERSTROM, R., 'Punishment', in *Philosophy and Social Issues: Five Studies*, ed. Wasserstrom, R. (Notre Dame, Ind., 1980), 112–51.

WILLIAMS, B., *Ethics and the Limits of Philosophy* (Cambridge, Mass., 1985).

WILSON, J. Q., *Thinking About Crime* (New York, 1975).

—— *Thinking About Crime* (revised edn.) (New York, 1983).

WOOD, DAVID, 'Dangerous Offenders and the Morality of Protective Sentencing', *Criminal Law Review* (1988), 424–33.

ZIMRING, F. E., and HAWKINS, G., *The Scale of Imprisonment* (Chicago, 1991).

Index